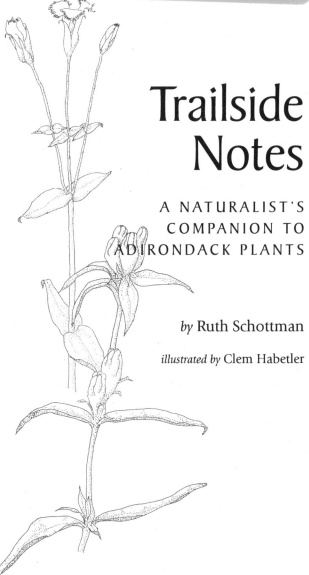

Trailside Notes

A NATURALIST'S COMPANION TO ADIRONDACK PLANTS

by Ruth Schottman

illustrated by Clem Habetler

Adirondack Mountain Club
Lake George, New York

This book was made possible with the generous and enthusiastic support of Gerhard and Evelyn Salinger. Additional funding was provided by the Schenectady Chapter of the Adirondack Mountain Club.

Earlier versions of these essays and many of the accompanying illustrations appeared in the popular Trailside Notes column in Adirondac, *the magazine of the Adirondack Mountain Club.*

Published by the Adirondack Mountain Club, Inc
814 Goggins Road, Lake George, New York 12845-4117
www.adk.org.

Design and typography by Michele Phillips Type & Design
Watercolor for cover by Mark Habetler based on drawing by Clem Habetler

Printed in the United States of America
04 03 02 01 00 99 98   10 9 8 7 6 5 4 3 2 1

Library of Congress Cataloging-in-Publication Data
Schottman, Ruth, 1927–
Trailside notes: a naturalist's companion to Adirondack plants / by Ruth
    Schottman; illustrated by Clem Habetler.
        p. cm.
Includes bibliographical references and index.
ISBN 0-935272-81-X (pbk.)
1. Botany—New York (State)—Adirondack Mountains.   I. Title.
QK177.S36   1998
581.9747'5—DC20
ISBN 0-935272-81-X

The Adirondack Mountain Club (ADK) is dedicated to the protection and responsible recreational use of the New York State Forest Preserve, parks, and other wild lands and waters. The Club, founded in 1922, is a member-directed organization committed to public service and stewardship. ADK employs a balanced approach to outdoor recreation, advocacy, environmental education and natural conservation.

Trailside Notes

Dedication

To family and friends who wandered and wondered with me.

Acknowledgments

C lem Habetler is more than the illustrator of these articles. I am grateful to her for a long and stimulating friendship, for her observational skills and her critical responses.

These articles have all appeared in *Adirondac*, the magazine of the Adirondack Mountain Club (ADK), although I have revised them for this collection. My association with ADK has been a happy one. After five children and many years in the Finger Lakes area of New York, I finally climbed an Adirondack high peak with illustrious companions—Ed Ketchledge, Orra Phelps and Eleanor Brown. I absorbed information and attitudes from them and felt at home at last, connected once more to mountains.

About this time, Nancy Slack encouraged me to teach her adult education class in natural history. I got hooked and have been teaching ever since. A little later, Orra Phelps retired and turned over to me her wild-flower classes at the Wildlife Federation's Summit at Silver Bay. There I met Barbara McMartin, then editor of *Adirondac*, who asked me to write my first Trailside article. Neal Burdick has encouraged me to continue. Thanks to all these nice people.

Contents

Introduction

Since early childhood, which I spent in the Alps of Austria, I have loved trees. When we became refugees, a tiny branch of a special tree came with me to the new world. I learned, many years later, that my tree was a walnut.

I seem to have had no botanical leanings until I entered college. At Cornell's College of Agriculture I wanted to learn how to produce more food for the needy post-war world. From studies in plant and animal breeding I gravitated to more basic science and majored in genetics. For fun, I enjoyed natural history field courses with Laurence Palmer and Eva Gordon. They required daily journal entries, which stimulated me to be more observant while en route between my classes, job and home. I avoided taxonomy, preferring instead to let names and relationships come naturally and gradually.

I have learned one general principle from observing nature and reading about it: there are many ways of coping with life. For some of us, nature watching is a release from the relentless consciousness of self. We find joy in empathy with other organisms. We become rooted amongst them. We can indulge in voyeurism without guilt. Rachel Carson wrote, "Those who dwell among the beauties and mysteries of the earth are never alone or weary of life," and Charles Darwin thought, "A traveler should be a botanist, for in all views plants form the chief embellishment." No matter how old you are, new discoveries—new to you—await you on every outing.

TIPS FOR THE FIELD

What will you need to explore and observe the world around you? Your senses and a receptive attitude are most important. Cultivate the friendship of fellow naturalists. More eyes discover more organisms, knowledge is pooled, speculations are tossed around and reference books can be shared!

Before you venture off the trail, where this is permissible, you should be familiar with poison ivy. Note the three leaflets, the central one on a longer stalk than the side ones, irregular teeth on the leaflets and alternate place-ment along the woody stem. The buds are pointed and covered with cinnamon-brown fuzz, but have no scales. The fruit is an ivory berry, growing in little clusters that hang on into winter. Poison ivy can be a low-growing ground cover or it can climb trees or fence posts, reaching out with horizontal branches. In a sensitized person an itchy, weepy rash develops a day or more after exposure to the plant itself or to clothes, tools or pets that have been exposed. A few other plants also can cause trouble, such as wet wild parsnip. Nettles cause almost instant pain, but it does not last long.

Poison Ivy

For various reasons, including protection from insects, ticks and sunburn, I wear long pants tucked into socks. Pencil and notebook are important to me, for shorthand notes and crude sketches. Some people prefer not to carry identification books. They leave them in the car or at home and consult them later, using their memory or notebook as a guide. If you adopt this method and realize you forgot to observe an important detail, you will do better next time.

You will undoubtedly want to acquire a library of identification books. For wildflowers in the Northeast, my students are about equally divided in their preference for Newcomb or Peterson (see Selected References). Most people record their observations directly in their field guides. Use waterproof ink. Some people even color the black and white pictures of new floral acquaintances, using colored pencils.

Many naturalists keep perennial calendars of events in the natural world, such as flower bud opening, fruit drop, full autumn color and so forth. Loose-leaf notebooks serve well for this purpose. I gather that with a computer and the right skills, you can enter and retrieve data by location, species, dates and events.

Because, as Rachel Carson says, "Some of nature's most exquisite handiwork is on a miniature scale," I highly recommend a 10x magnifying lens, worn on a string around your neck. For advice on good lenses, binoculars, waterproof notebooks, compass and maps, knapsacks, clothes and field techniques, consult David Pepi's book (see Selected References). I also carry plastic bags in my pocket, as well as some paper towels and a length of surveyor's tape. The plastic bags are handy for kneeling or sitting on wet ground, keeping books dry and possibly bringing back a small sample of a non-scarce plant; the paper towels are good for wiping mud from your hands before consulting a book. I use surveyor's tape for marking a spot, such as my entrance point to a bog, as I usually want to return the way I came.

As for plant presses or other ways to preserve plant material, I rarely collect anything but leaves. They can be labeled very nicely with magic markers and pressed in a coloring book or phone book. Twist-ties are useful for marking a particular plant or part of a plant you wish to watch over time. Maybe little mementos from the field, along with your notes, will help you relive your outings and etch them in your memory.

Compass, maps and knowing how to use them are important if you venture into the unknown. Consult Bruce Wadsworth's *An Adirondack Sampler II: Backpacking Trips for All Seasons* or lists provided by the Adirondack Mountain Club for recommended supplies for longer outings.

Many of us are enthusiastic hikers as well as naturalists. Therein lies a dilemma. Don't strain the tolerance of your hiking companions by indulging in too much botanizing. While I search for a plant or work on an identification, my husband likes to photograph or nap or search for mushrooms, so it is possible to explore together sometimes.

BETWEEN FIELD TRIPS

There are days when searching in the library might be preferable to exploring outdoors. Hunting along the "500" shelves—especially those whose books bear call numbers in the 570s and 580s—can be productive and serendipitous. Don't forget the children's section. Great photographic essays often find a home there. I have found much pleasure and inspiration from Thoreau, Darwin, John Burroughs, Edwin Way Teale, Hal Borland and contemporary writers such as Hubbel, Stokes, Russell and Serrao (see Selected References). We all, well-known naturalist–teachers and amateurs, do essentially the same thing. We combine our limited observations with research through books and journals and synthesize. Given the genetic diversity within each species and diverse environmental conditions, multiple observations are important and always valid.

The essays in this book are about common plants, those I could observe year-round and those you are likely to see. I have omitted the fascinating residents of mountaintops and bogs, both of which are important Adirondack habitats. Nancy Slack and Allison Bell have described the former very well in a beautifully illustrated book, *85 Acres: A Field Guide to the Adirondack Alpine Summits*. You will find good bog plant descriptions in Charles Johnson's *Bogs of the Northeast*.

These essays are written for the curious amateur and are arranged for seasonal browsing, beginning in early spring. I hope you will be stimulated to go and look for yourself. Perhaps you can publish your nature observations in a local newsletter or newspaper. If you want company in the field, check at nature centers, museums, environmental education centers and arboretums. Try workshops offered by ADK and others. Attend natural history conferences at the New York State Museum; join the New York State Museum's Flora Society.

Parts of a Flower: A Visual Primer

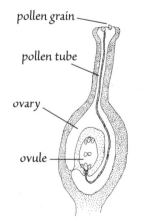

Pollen is usually brought to a pistil from another flower on the bodies of insects or birds, or by the wind. A slender tube grows from the pollen grain down through the pistil to the ovule where the sperm cell fertilizes the egg. While still in the tube, the sperm cell divides and makes two

pollen grain

pollen tube

ovary

ovule

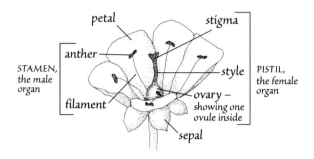

STAMEN, the male organ

PISTIL, the female organ

petal

stigma

anther

style

filament

ovary – showing one ovule inside

sepal

cells. The second sperm cell joins two other cells in the ovule to make the endosperm, a nutritive tissue that will grow and feed the developing embryo.

Now the ovule can grow into a seed. The ovary becomes the fruit.

Spring

Human snowbirds return north. Hibernators emerge. Even those who have enjoyed the outdoors in winter greet signs of spring with joy. No other season brings as many novices to natural history classes as spring.

This is the time for exploring our woodlands. Most herbaceous plants bloom before the leaf canopy closes over them. Some ephemerals rush their food-making and storing activities and their flowering and fruiting. Then they disappear from the surface of the forest by summer. Others allow us to watch their development at a more leisurely pace.

Don't ignore woodland edges where trees and shrubs get enough sun to maintain low branches. This is the best place for watching the opening of buds and the unfolding of branches bearing leaves, or flowers, or both.

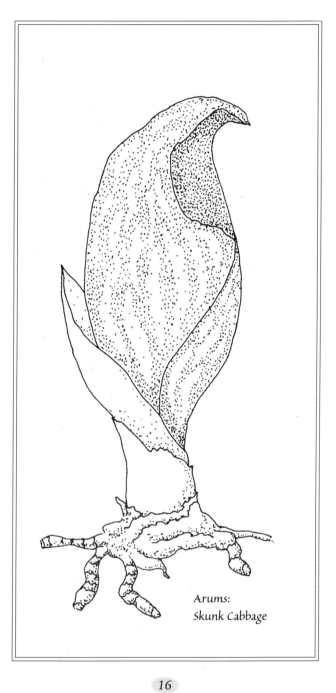

Arums:
Skunk Cabbage

The Fascinating Arums: Skunk Cabbage

M ost people are acquainted with skunk cabbage *(Symplocarpus foetidus)* and jack-in-the-pulpit or Indian turnip *(Arisaema triphyllum)*. If your relationship has been a distant one so far, get a little closer and enjoy the fascinating aspects of these plants. They are cousins in the Arum family, a family whose members usually bear a spadix and spathe. A specialized hooded leaf (the spathe) enfolds a club-like structure (the spadix) in which many small flowers are embedded. Both species like moisture.

Skunk cabbage enjoys a very wet habitat. You will find it in the lowlands surrounding the Adirondacks and all the way from Nova Scotia to North Carolina and west to Minnesota. It blooms before its leaves unfold. You may have seen some spathes already poking out of the ground in late fall. More appear in late winter and early spring when the plants begin to bloom. How do the flower tissues avoid being frozen?

In an article published in *Science* (Nov. 22, 1974), Roger Knutson reported on his skunk cabbage research, which involved taking spadix temperatures and measuring oxygen consumption in the field. He found that the spadix maintains a flower temperature of about seventy-two degrees Fahrenheit for at least two weeks. Knutson says heating is most intense during the early stage of flowering. At this stage, the female organs are receptive; pollen is not yet being shed. There are many plants whose male and female organs are not ready for sex at the same time. This, of course, prevents self-fertilization and also points out the need for an agent to carry pollen from one plant to another.

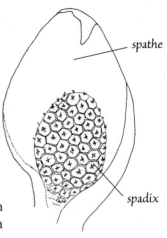

spathe

spadix

When you peek into the spathe of blooming skunk cabbage you may see a pollinator, or just a guest. Bees have been seen making a rest and recovery stop on the spadix. They warm up and regain their flight ability in weather below fifty degrees Fahrenheit. To keep its temperature up, the spadix respires rapidly, consuming oxygen just like an animal of similar size. The colder the day, the faster the spadix respires. Carrying just a spadix in your pocket will not warm your hands for long, though it will make them smell quite skunky.

The roots of skunk cabbage supply energy in the form of starch. The roots anchor the plant, which, Knutson says, may live for *many* years. Every spring they wrinkle in unison, pulling the stem down a distance equal to its annual growth in length. Ginseng performs the same trick and shows annual wrinkles on its neck. I suppose all nonwoody perennials have to accommodate to growth somehow or find themselves lifted out of the soil. Checking on the root wrinkles of skunk cabbage requires a hardy investigator.

Jack-in-the-Pulpit

O ur second Arum is jack-in-the-pulpit (*Arisaema triphyllum*), also known as Indian turnip. Although it is visible later in spring than the subjects of the following essays, I include it here for proximity with its relative, the skunk cabbage. The underground storage organ (the turnip) can only be eaten after laborious treatment. Better leave the turnips in the moist forest soil and become a jack-watcher.

The spathe may be solid green or striped with white or purple. At the base of the spadix are small male or female flowers. A male flower consists of white or purplish anthers that release pollen, which collects at the base of the spadix in a chamber made by the surrounding spathe. Female flowers look like green peas topped by a receptive stigma. Each contains several ovules that may become seeds if fertilized.

Fungus gnats serve as carriers of pollen from plant to plant. What attracts gnats to a jack is not known. Once inside the spathe, the gnat has trouble flying up and out again, or crawling up the slippery wall. The life of the gnat seems doomed, but pollination has not been achieved.

This puzzled me for years until I read in an article, "Jack and Jill in the Pulpit" by Bierzy-Chudek (*Natural History*, March 1982), that while the spathe of a female plant overlaps neatly at its base, there is a little escape gap at the base of spathes in male plants. Bierzy-Chudek's descriptor line beneath the title reads,

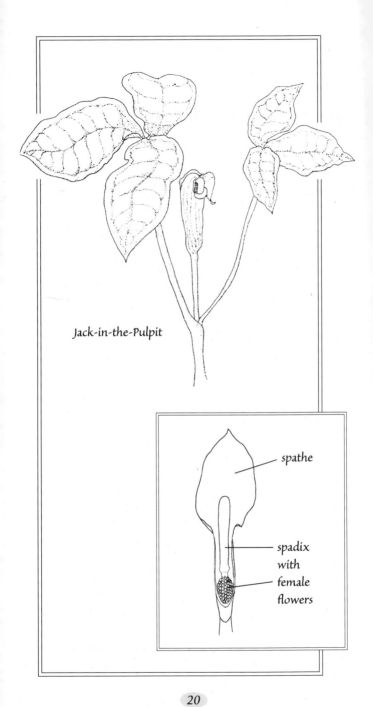

Jack-in-the-Pulpit

spathe

spadix
with
female
flowers

"For some small plants, the burden of being female is just too much to bear."

She verifies and quantifies the facts that we jack-watchers have long known: very small one-leaf jacks bear no flowers; somewhat larger plants bear male flowers; large two-leaved plants bear female flowers. She does not mention the possibility of male flowers subtended by female flowers, which I have also found.

Sex is determined the previous summer, and depends on the nutritional status of the plant. Sex is reversible, an adaptation to the environment. Sex changes occur in many species of animals, such as tropical fish, but seem rare in plants. Presumably the production of large seeds by jack (jill, I mean) is an energy-draining effort for a perennial that must also store food in its corm for its own future.

You can verify others' observations and carry on your own research. As you approach a jack (jill?), guess its sex from its size. Approach closer. Is there a gap at the base of the spathe? Can you peek inside carefully? Was your guess correct? Are there any captive gnats?

By late summer look for red ripe berries on an old spadix that has burst out of its spathe. Bierzy-Chudek found an average of less than ten seeds per plant per year on her upstate New York study site. (I have consistently counted more.) The author got better results when she artificially hand-pollinated flowers with a camel's hair brush. The problem of poor pollination, the evolution of a relationship where the successful pollinator languishes in prison rather than being freed to breed more successful pollinators, is an interesting philosophical one. Is it really a problem to jack, or do you find that this plant reproduces quite successfully?

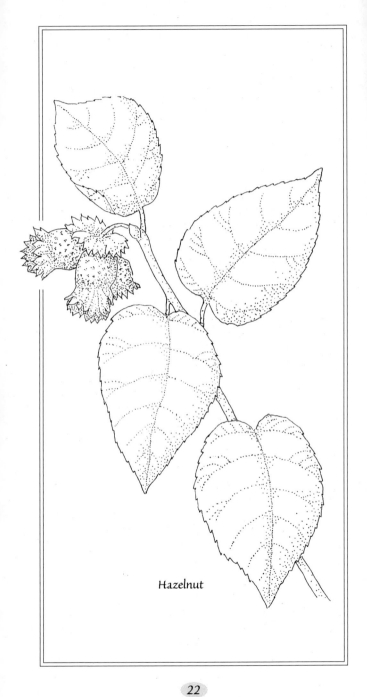

Hazelnut

Hazelnut

Earliest to bloom among our woody plants is hazelnut, also known as filbert. To see it, you need an observant eye. Look for a shrub in open woods, at the edge of a forest or in thickets. The shrubs may have catkins, unisexual spikes of flowers with scaly bracts. Ever since summer, catkins bearing the male flowers have been visible on young branches, but usually not at the tip.

Winter catkins are not common. Birches have winter catkins, at the tips of branches, as does another member of the birch family—hop hornbeam. By the time these two woody plants reach sexual maturity and make catkins, they are trees. Among shrubs, alders show catkins of two sizes (one for each sex), but they grow in wetter habitats than hazelnut and are often decorated with fruit cones. Sweet fern and sweet gale are tip-of-branch catkin bearers, but small, less than five feet. Hazelnut's catkins are often deformed into a candy cane shape.

In late winter or early spring, while hazelnut catkins are still tight, brilliant red stigmas protrude from the scaly buds near branch tips. These are the receptive parts of four to sixteen female flowers waiting to be pollinated by windblown pollen. Eventually the catkins stretch out, hang loose and offer ripe pollen to the wind. Once, in late March, I shook such a catkin, caught a load of pollen in my hand and apparently rubbed my eyes sometime later. I soon had swelling, aching, itching eyes. At the emergency room they said

I had a "typical allergic reaction." Don't rub pollen in your eyes.

Below the receptive stigmas there is as yet no ovary with eggs waiting for sperm brought by the pollen. Within four to seven days after pollination, the pollen tube grows to the base of the female flower and goes into "hibernation." Now the ovary is stimulated to develop. By mid-June the eggs are ready and the resting sperm becomes activated. Fertilization occurs several months after pollination. A similar delay occurs in the fall-flowering witch-hazel. (Some animals also arrest their reproductive process at various stages.)

The bracts that surround the female flower enlarge to surround the fruit. They remain mostly separate in the American hazelnut; they join to make a very long, narrow tube in the beaked hazelnut.

Without fruit, you can usually tell our two native hazelnuts apart by the number and appearance of hairs on young branches and leaf stems. American hazelnut (*Corylus americana*) has both nonglandular and gland-tipped hairs, lots of them. Beaked hazelnut (*Corylus cornuta*) has few hairs, all nonglandular. A glandular hair looks like a bristle with a spherical sticky blob at the top. Use a hand lens to examine them.

The ranges of the two species overlap; the beaked hazelnut has a more northern affinity. The leaves are similar: coarse, double-toothed and veiny. They do not distinguish themselves with brilliant fall color.

By fall the fruit has ripened. The husks surrounding

the nuts dry and spread open, and the nuts fall to the ground. Usually squirrels, other rodents, birds and deer eat the fruit, even before it is ripe. There are few left for people and fewer for reproduction.

As with other nut-bearers, there are exceptional years now and then when hazelnut outdoes itself; 1996 was such a year in Albany's Pine Bush. In July the bushes were bending over under the weight of the fruit clusters. There should have been enough for the usual consumers and leftovers for starting another generation of shrubs from seed. Most years all reproduction is by shoots from the roots.

The cultivation of hazelnuts has been going on for over 4,500 years. It is the European hazelnut (*Corylus avellana*) that is the basis of the American commercial venture, which is located almost entirely in Oregon. Cultivated American hazelnuts and crosses with European hazelnut have been tried in noncommercial situations in New York with variable success. If you are successful with peaches, try hazelnuts.

I grew up in Austria, where hazelnuts, often coupled with chocolate, appeared in many sweet guises. Oregon hazelnut growers encourage you to try them in everything from soup to nuts.[1]

Traditionally, hazelnut branches were used for divining water or precious minerals. Charcoal from the wood was used for drawing. I have read in the journal of the North American Mycological Association that Perigord truffles, the diamonds among truffles, can now be grown by inoculating young hazelnuts.[2] "Seeking the *Truffe*," another article about truffles (*Natural History*, January 1996), also references Garland (footnote [2]).

1. A good source of information is Hazelnut Growers of Oregon, 401 N. 26 Avenue, P.O. Box 626, Cornelius, Oregon 97113.

2. For information, contact Garland Gourmet Mushrooms and Truffles, Inc., 3020 Ode Turner Rd., Hillsborough, North Carolina 27278.

Son-before-Father:
Coltsfoot

Dandelion

Son-before-Father: Coltsfoot

As snow cover and bare ground alternate in early spring, we begin our search for the first flowers of the year. If we look up we may see elms and silver maples in bloom. At eye level, we are likely to spot pussy willows discarding their dark bud scales and alder and hazelnut stretching out their catkins in preparation for releasing pollen. In wet places in the lowlands around the Adirondacks, skunk cabbage is blooming, but it is often missed because it is dark-colored and not so accessible.

On sunny days there is a bright yellow flower greeting us from roadsides, ditches and disturbed slopes. From a distance you might mistake it for dandelion, but upon closer inspection you'll find it is coltsfoot (*Tussilago farfara*). Its blooming season is long, though our perception of it is limited because the plant keeps its flowers closed and protected on cloudy, rainy days. We get a chance to see buds, flowers, closed flowers ripening and ripe flower heads dispersing fruits, all at the same time, by mid-May.

Coltsfoot's name derives from the shape of its leaf. You will have to judge whether this is an appropriate name after most of the blooming is done. *Then* the leaf buds open. Thus the fanciful English name, son-before-father. The leaves, which are covered with white fuzz on the underside, gradually enlarge, sometimes reaching seven inches across.

This perennial plant is from Europe, where it is held in high esteem. It has become widespread in northeastern America and is common in the lower Adirondacks. Its thick, branched, spreading rootstock (rhizome) helps to hold soil on slopes. It was used to make a dye that gave wool a greenish-yellow color. The leaf shape appeared on apothecary signs, signifying its medicinal use. Euell Gibbons waxes rhapsodic, in his *Stalking the Healthful Herbs*, about the flavor of products made from coltsfoot.

Cough drops and syrup are made from fresh leaves in June or July. Coltsfoot tea is made from dried leaves. You can also smoke dried, crushed coltsfoot leaves. Gibbons tells us further that "Greeks and Romans treated asthma by burning this herb on charcoal fires and inhaling smoke through reeds, alternating puffs with sips of wine." According to Adrienne Crowhurts's *The Weed Cookbook*, it was the *rhizomes* of coltsfoot that were used for making medicine. If I had some coltsfoot on our property I would be tempted to try her simple cough drop recipe. Coltsfoot's generic name, *Tussilago*, comes from the Greek word for cough: *tussis*.

This is a wonderful plant for teaching purposes, an immigrant sufficiently aggressive to be considered a weed. So I can pluck a head and distribute pieces of it into waiting palms. Here we have a member of the Composite family (see White Snakeroot for a discussion of Composites). Each head is a bouquet of flowers and everyone has several flowers in hand to examine with a hand lens.

If a dandelion can be found nearby, it is fun to compare the two yellow-flowered composite cousins. Even young children can point out the differences in leaves and stems. Can you blow through a coltsfoot stem? The flower head differences are more subtle. Whereas all flowers on a dandelion head are strap-shaped and bear both male and female organs, enabling all flowers to make fruits, coltsfoot's head is more complex. There are several rows of strap-shaped flowers on the outside, equipped only with pistils. The center flowers are tube-shaped. These flowers have functioning stamens but only rudimentary pistils, so they cannot produce fruits. (Botanists call them sterile even though they furnish the pollen necessary for fruit development.) Only the outer circles of flowers will bear fruit. (The seed enclosed by the ripened ovary wall is called a fruit, if you wish to be botanically correct.) Compare the fruits of coltsfoot and dandelion. Dandelion fruits sail through the air on a stalked parachute. Are coltsfoot fruits missing something?

A patient friend watched a coltsfoot flower head open one morning. It took ninety minutes. The head was one and one-half inches across when fully expanded. It reopened most days for a week. Each day the head turned with the sun. Eventually the head remained closed, in a nodding position. Flowers changed to fruits while the stalk lengthened. Finally the climactic day arrived. The stem straightened and the bracts unfolded and presented a fluffy head to the breeze.

I would love to know the relationship of coltsfoot to animals other than humans. Who are its pollinators? Who feeds on its leaves, rhizomes, fruits? Perhaps next spring will bring some answers.

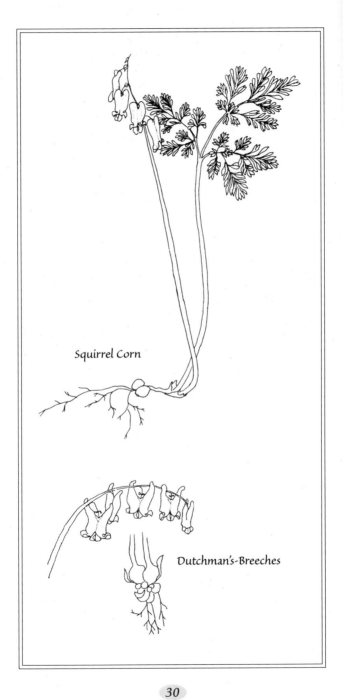

Squirrel Corn

Dutchman's-Breeches

Lacy Cousins: Dutchman's-Breeches and Squirrel Corn

In early spring you will surely meet one or the other of the lacy cousins, Dutchman's-breeches (*Dicentra cucullaria*) and squirrel corn (*Dicentra canadensis*). I especially remember rich displays on the trail up Thomas Cole Mountain in the Catskills and on the Pitchoff trail in the Adirondacks. Some Dutchman's-breeches grow in my backyard, where I can more easily follow their lives. I keep a loose-leaf notebook calendar for recording observations: April 5: "Plants up, buds visible."

These plants are perennials. According to naturalist and educator E.H. Ketchledge, they are indicator species for rich soils, especially where limestone is present. The delicate fern-like leaves, repeatedly divided by three and very tedious to draw, make food while the sun shines in early spring, before trees are in foliage. The food supports flowers, fruits and under-ground storage and reproductive structures. May 25: "Wolf Hollow: On the moss-covered talus below the dolomite cliff the foliage of both Dutchman's-breeches and squirrel corn has yellowed and wilted. The pale pink-gray crowded tubers of the latter are lying on the

D. cucullaria *Flower Parts*

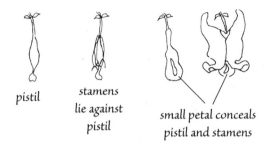

pistil

*stamens
lie against
pistil*

*small petal conceals
pistil and stamens*

surface, attached to dislodged plants."

First bloom of Dutchman's-breeches near Schenectady, New York, has occurred between April 10 and May 9. Squirrel corn usually blooms one or two weeks later. I sacrificed one flower for dissection. The two *Dicentra* flowers are very similar in construction. Note the two scale-like sepals. There are also real bracts on the flower stem at some distance from the other flower parts. The petals are of two kinds: two large, spurred ones and two inner, paddle-shaped ones that are united at the apex, forming a crest over the sex organs. These consist of six stamens in two groups of three and one small stigma on a slender style. "To my surprise the stigma of my dissected flower is full of pollen even though the flower just opened. I haven't seen any bees yet. Could the flower self-pollinate?"

April 30: "Insects have been chewing circular holes through the big petals to steal nectar." Books tell me that long-tongued bumblebees are the best pollinators of Dutchman's-breeches, but the bee must enter the flower properly to deliver pollen to the stigma and pick up another load.

May 6: "The majority of flowers, open three to five days, are punctured. A plant could have become extinct due to the misbehavior of insects if it did not have a vegetative means of reproducing." There is also a chance that self-pollination can be successful (experimenting by putting an insect excluder around a flower tempts me) and we have already met the tubers by which means *Dicentras* reproduce asexually. Squirrel corn flowers are supposed to have a pollinator-attracting odor, a smell like that of hyacinth. At noon, a group of us smelled nothing. Perhaps at a different hour one's nose would be rewarded.

If pollination is successful, the single ovary may become a two-valved capsule with ten to twenty shiny

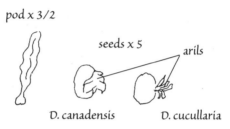

pod x 3/2

seeds x 5

arils

D. canadensis   D. cucullaria

little seeds. Each seed has a nutritious appendage, called an aril, that attracts ants. An ant is likely to drag a seed to its home, eat the aril and "plant" the unharmed seed. We are learning about more and more woodland flowers that depend on ants for their dispersal.

Dicentras are members of the Fumitory family. Other members of this family that you might meet are *Corydalis sempervirens* (pale corydalis) and *Adlumia fungosa* (Allegheny vine). It is best not to eat any Fumitories as they contain alkaloids that have poisoned livestock.

Bloodroot

Bloodroot

Bloodroot (*Sanguinaria canadensis*) produces one of our most admired early spring flowers, but you need to be out on a sunny day to appreciate it fully. On dull and rainy days the petals close. Wind and rain may shorten the season of bloom to a day or two.

Bloodroot grows throughout most of eastern North America. The best place to look for it is in a rich (meaning not acid), well-drained, open woodsy setting. To see the two spoon-shaped, off-white sepals you need to be around when the flowers open or peek at a flower bud before it blooms. The bud is enveloped in a single leaf, with prominent veins on the underside, and the leaf is wrapped in papery bracts.

Like many members of the poppy family, the flower sheds its sepals as the soft, sparkling white petals unfold. There are usually eight petals, in two series, but the number may vary from six to twelve. My bloodroot had twenty-four golden anthered stamens, but that number also varies. There is just one pistil.

I watched a fly among the stamens. There is pollen available for all kinds of bees, but no nectar. After pollination, the flower stalk grows taller and the leaf grows much bigger, usually winning the height competition.

In 1990, my companions watched a number of seed pods (ripe ovaries). Ripe pods opened between June 13 and 18. One described the event: "Both sides split lengthwise, leaving a central structure like the eye of a large needle. Most of the numerous seeds escaped." My companions gave me a pod nearly two inches long with plump shiny brown seeds, about one-eighth inch long. A white crest, like a delicately patterned slug, wrapped around the seed. This is an aril, or elaiosome, a special nutritious and appeal-ing organ that attracts ants, which haul the seeds to their nests, eat the arils, and "plant" the unharmed seed. Research shows that this ant habit has reduced seed loss due to rodent predation from 84 percent to 43 percent.

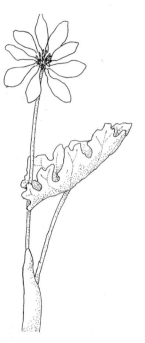

Forty years ago, in a more ignorant era, when we more freely multiplied ourselves and decimated plants, I dug to the shallow rhizome (underground stem) and cut off a piece to demonstrate the reason for this plant's common and scientific name. It "bled" profusely. I even applied bloodroot's dye to eager hands and faces. Today I practice restraint.

In *A Weaver's Garden*, Rita Buchanan writes,

> Indians used bloodroot as a body paint, and since colonial days dyers have tried to capture that pigment on textiles. The bright colored sap is nature's way of saying keep away. It is quite toxic if ingested, causing vomiting, dizziness, nerve damage, and even death. No dyer is going to eat bloodroot; it tastes awful, anyway. I dyed some wool fleece with bloodroot, rinsed it carefully and let it dry. Days later I began to card it for spinning.

Within seconds, my nose and throat were inflamed, my face got red, and my head felt like it was going to burst. I stumbled into the next room, got out my book on poisonous plants and read that bloodroot sap is a potent irritant of the moist membranes. I had inhaled bits of dried sap that came off the fibers in the carding process.

Bloodroot's dye has inspired common names such as Indian paint and red puccoon and its medicinal use gives us "snake bite" and "sweet slumber."

The deep orange-red color of the latex, found throughout the plant but concentrated in the rhizome, is caused by the alkaloid sanguinarine. A study reported in *Rhodora* (April 1990) was concerned with sanguinarine amounts in dried rhizomes of bloodroot growing in various habitats. Content varied from .6 percent to 6.3 percent, a good fact to keep in mind if one is tempted to use natural medicines.

sepal

At very low concentrations this substance inhibits growth of the fungus responsible for root rot. Sanguinarine has been used experimentally for ulcers, ringworm and skin cancer. In 1983 we noticed it in an anti-plaque toothpaste. I called the manufacturer's 800 number and was told extract of sanguinaria is still used in the toothpaste but the amount is proprietary information and that it is obtained from an herb grower who "plants the root again after the extraction." I wonder.

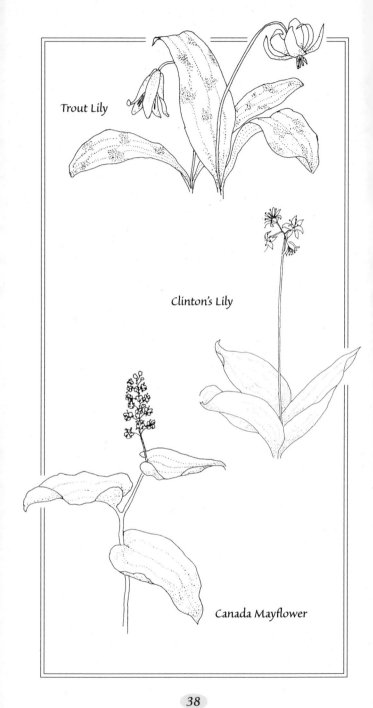

Trout Lily

Clinton's Lily

Canada Mayflower

Lilies

On your walks along Adirondack trails you are likely to meet at least ten members of the lily family. We'll take a look at three common ones—trout lily, clintonia and Canada mayflower.

Trout lily (*Erythronium americanum*) is one of our earliest spring flowers. You'll find it up to elevations of 2500 feet. The fruits mature and the leaves wither by the time the hardwood trees are in full leaf. According to an article in *Science* (September 17, 1976), the above-ground activity of this plant extends from snowmelt to the development of leaves in the forest canopy. The plants help reduce losses of potassium and nitrogen from the ecosystem during the periods of maximum snowmelt and runoff by incorporating them into their own biomass. Later, during summer, these nutrients are made available to other plants when the above-ground tissues of trout lily decay. The speed of recycling is impressive. One writer referred to the decay of leaves as self-digestion.

The three petals and three sepals, all yellow, form a bell-shaped flower. You can tell which are the three sepals by their outer position and slightly different shape, texture and coloring. On dark days and at night this flower "goes to sleep," drooping, probably protecting the stamens and pistil from wetness. On bright days petals and sepals are reflexed. Inside the bell is a clapper of six stamens. Insects cling to the stamens when probing upwards for nectar. The green pistil has a long style that remains after other flower parts have dropped. The fruit is a somewhat triangular capsule that splits into three sections and sheds many crescent-shaped seeds. If you wonder why this yellow-flowered plant is named *Erythronium* (*erythros* means red, in Greek), it is because its European cousin is reddish-purple flowered.

At the base of a trout lily stem, at consider-able depth in the soil, is the expanded food-storing organ known as a corm. Apparently this white corm is the reason for its other common name, dogtooth violet. Why was "violet" added as a surname? Maybe it was due to the similarity in the fruits. The trout lily can reproduce vegetatively by new corms formed from the old corm. So the books say, without a detailed description. I have not been in a situation where I could thrust a spade deeply enough in a patch of trout lilies to see what goes on around the corms. But for years I have wondered about ghostly white shoots arching out of the soil and sticking out of the sides of eroded trails. These are most apparent after the trout lily plants have disappeared above ground. Are new corms formed at the end of these organs? Apparently, yes. New corms and seeds send up single-leaved plants for three to six years until they gain enough food reserves to produce a mature two-leafer with flowers. Some colonies of single-leaved plants remain that way for many, many years. They never flower and only reproduce asexually by the dropper and corm method. A sexless variety seems to have evolved.

The leaves, four to ten inches long, are mottled brown and purple. John Burroughs, the Catskill poet-naturalist, is responsible for giving the plant the names fawn lily and trout lily. Adder's tongue is another interesting name for this beautiful plant.

Clinton's lily, *Clintonia borealis*, was named for New

SPRING

York's Governor DeWitt Clinton (1769–1828), who was also a naturalist. It is very common in the Adirondacks up to timberline. You can climb mountains, observing clintonia with ripe, many-seeded blue berries at the base, with green fruits higher up, with flowers above that, and maybe with buds near treeline. This lily bears yellow-green bell-shaped flowers nearly one inch long, with three to six flowers on one stem. Petals and sepals are nearly identical. Shiny leaves, much longer lived than trout lily leaves, usually come three per plant, but vary from two to five. The names cow's tongue and bear's tongue undoubtedly come from the leaves. Bead lily or blue-bead obviously come from the fruit. But why dog-berry? The berry is considered poisonous for humans. Was it used to poison dogs? The plant can reproduce vegetatively from branches of its slender rootstock. I don't know how long it takes a new plant to mature.

The characteristics of the lily family include perennial habit, parallel-veined leaves, and a flower plan of three. Trout lily and clintonia are conformists; false lily of the valley (*Maianthemum canadense*) is a non-conformist. This small plant, also called Canada mayflower, often carpets large areas of the forest and climbs up to timberline. *Maianthemum* means *may flower* in Greek. The leaves are parallel-veined, about three inches long, heart-shaped at the base, and pointed at the tip, with one to three leaves attached to a zigzag stem. Young, prepubescent plants are one-leafers.

To see the details of the half dozen to dozen flowers at the top of the stem you'll need a magnifying glass. You will be surprised to note that this lily has a perianth (petals and sepals) of four parts and four stamens, a very unusual feature for lilies, thus the term "nonconformist." The flowers become white berries with spots, later turning pale red. Bead ruby is another name for this plant. Are the berries poisonous? I checked in Hardin and Arena's *Human Poisoning from Native and Cultivated Plants*. They say, "Caution!" This plant, too, has a vegetative reproductive strategy, by means of its rootstock.

Goldthread

Laurence Palmer, in his *Fieldbook of Natural History*, finishes his text on goldthread (*Coptis trifolia*) with "whole plant—a thing of beauty." I am sure you will agree.

You will find the plant in moist areas up to timberline. It loves bogs, shady areas at the base of conifers and beds of moss. The leaves are all basal and consist of three leaflets. The leaflets are prominently veined, dark green and shiny above, sharply toothed, obscurely lobed and evergreen. The generic name *Coptis* is derived from the Greek *kopteion*, to cut, referring to the divided leaves. The specific name tells you the number of leaflets! In the spring, new leaves uncoil from a golden bud to replace the "evergreen" ones.

One or several flower stalks grow from the plant, bearing one flower each. If you look carefully, you will see a scale on the stalk, a little beneath the flower. The flower, almost one inch in diameter, has five to seven petal-like sepals that drop off easily. Next there are an equal number of petals, or staminodes, depending on which authority you consult. These organs, by either name, have glistening nectaries on their tips.

There are numerous stamens with golden anthers and three to nine pistils. The ovary of each pistil is elevated on a two-millimeter stalk. After pollination (fungus gnats and small beetles are the pollinating agents, according to Dr. Palmer, but you might find other insects), the stalks elongate and the ovaries

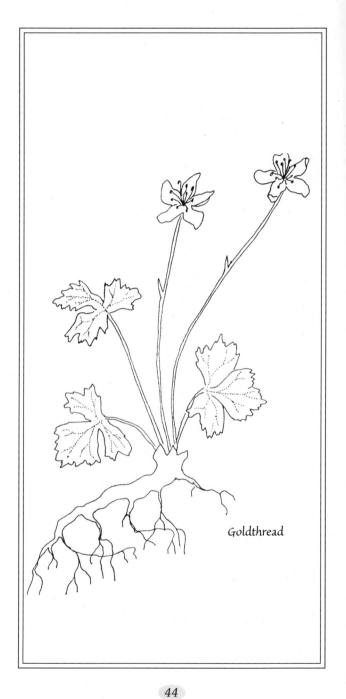

Goldthread

become fruits, called follicles. Each follicle contains four to eight seeds. The umbel of follicles persists for months.

The word *gold* has been mentioned several times, but no thread. Poke your finger into the moss and soil at the base of the plant and you will expose the rhizome (rootstock), which is truly gold in color. It is also very bitter to taste. Was its medicinal use the consequence of its bitterness? Rhizome and roots, whole or powdered, were used for healing sores of the mouth and eyes, for thrush, for a tonic, as a cure for alcoholism and as a dye; thus we have other names for goldthread such as canker-root and mouth-root.

Wild Ginger

Wild Ginger

Wild ginger (*Asarum canadense*), one of my favorite woodland wildflowers, blooms in April and May, but the flower persists for many weeks, gradually maturing to fruit. The foliage is handsome and long-lasting, but not evergreen like that of a related species that lives in the Northwest. Each spring, preformed buds open on the perennial root-stock, or rhizome. From each bud come paired leaves and, between them, a solitary red-brown flower, an easily missed beauty.

John Burroughs wrote, "Why should this plant always hide its flower? As a rule the one thing that a plant is anxious to show, to flaunt before the world, is its flower." The flower is often half-hidden in duff. You will have to get on your belly or knees or observe from the bottom of a slope or limestone ledge.

The pistil matures before the stamens, a device used by many plants to discourage self-fertilization. Twelve stamens are attached to the top of the ovary, which is therefore referred to as inferior. It holds many ovules, which may become seeds, in six compartments. There are no petals, the usual showy parts of flowers. The sex organs are surrounded by a reddish calyx with three flaring lobes. Flies are attracted to the flowers and serve as pollinators.

"The rootstock is not offensive to taste," says Laurence Palmer in his *Fieldbook of Natural History*. I prefer just to smell the wonderful gingery odor. Cut a small part of an older rhizome that has no bud.

The Chippewa people used wild ginger as medicine for indigestion. The family name, Aristolochidaceae, or birthwort, suggests medicinal use for inducing labor. The ginger of commerce comes from an unrelated Asian plant, *Zingiber officionale*. One March I lifted some duff and looked at the network of rhizomes beneath a crowded bed of wild ginger. To my surprise, the rhizomes were not at all covered by soil (thus, they were really runners or stolons) and were green, indicating their ability to photosynthesize in early spring. The bits of rhizome that I had harvested before were always white. Was the green unusual, due to crowded conditions and lack of cover? Was it temporary?

I followed the three branches of one plant's rhizome and counted seven buds, each about one inch long. I took one bud to dissect. There were two outer protective bracts, two somewhat dilapidated leaves and blackened tissue instead of the promise of a flower between them. Maybe the exposure endured during winter was responsible.

The plant I examined grows confined by rocks in the illustrator's garden. Every spring the old leaves are raked away, leaving a much lighter covering of duff than would be normal in wild conditions. I suggest both you and I carefully stir around wild ginger to see how it lives.

I did learn something. There are not "two leaves per plant," as my book said, but two leaves per stem. (More than one branch may spring from one root.) I still don't know when the ripe fruit capsule bursts and I have never seen the seeds of wild ginger. Maybe you could join me in that next exploration.

Mayapple

One of the delights of spring is the annual emergence of the stout, ribbed stalk of mayapple (*Podophyllum peltatum*), each stalk wrapped in a leaf in the manner of a folded umbrella, and the ensuing unfurling of the umbrellas. Look for the plant in open, moist woods, along roadsides and forest edges.

Note that some plants have just one leaf, others two. (Occasionally there are three leaves.) The second leaf, a bit smaller, tends to have fewer lobes and a thinner stem. One-leafers do not bear flowers; two-leafers do. The die was cast in July of the previous year when buds differentiated on the stout, perennial, horizontal underground stem—the rhizome.

Here is Anna Comstock's description, from *Handbook of Nature Study*, of what you may watch in April and May. "The bud pushes its head out from between the two folded parasols and takes a look at the world before it is covered by its green sunshades."

As the bud unfolds it quickly sheds its six (or three) green sepals. Six (and sometimes up to nine) white petals spread out, and you can see a mass of yellow stamens grouped around the pistil. You probably noticed the great variability in numbers of flower parts mentioned. You may find variations other than those described. Each stamen has a grooved anther. The groove slits open and sheds pollen on the side away from the pistil. Pollen, but no nectar, rewards visiting bumblebees and other insects. After pollination, the

Mayapple

"Umbrella" opening

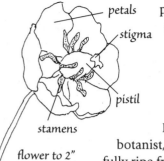

petals

stigma

pistil

stamens

flower to 2"

pistil with its ruffled stigma grows into a fruit. When ripe, this fruit is a fragrant, yellow berry about two inches long, with juicy pulp and many small seeds inside.

Richard Mitchell, our state botanist, says, "The pulpy flesh of fully ripe fruit is reportedly edible."

"They are not tasty," he says elsewhere. Asa Gray, famous botanist of a previous era, describes them as "mawkish, eaten by pigs and boys." On the other hand, James Whitcomb Riley in "Rhymes of Childhood" sings about the mayapple fruit,

> And will any poet sing of a lusher, richer thing
> Than a ripe mayapple rolled like a pulpy lump
>   of gold
> Under thumb and finger tips and poured molten
>   through the lips?

There is a tale in Oliver Perry Medsger's *Edible Wild Plants* about his consumption of a hatful of mayapple fruits, his severe belly ache a half-hour later and complete recovery two hours after that. DeWitt Clinton, New York's governor and a naturalist, wrote of the proposed sale of ripe mayapple fruit in New York markets after the opening of the Erie Canal. "Clinton's Ditch" became a reality, but the mayapple dream did not materialize.

Euell Gibbons has a mayapple marmalade recipe which, according to him, produces a unique gourmet product. I have followed his directions to the satisfaction of tasters. Most years I can't find enough ripe, unbitten fruits at the same time in August and September. I have learned to harvest them just before full maturity, let them ripen

fruit to 2" long

indoors and wait in the freezer till enough accumulate for jam or sauce. Mitchell and other sources warn us that unripe fruits, leaves and rhizomes of mayapple are not only poisonous to eat, but also may cause dermatitis when handled. Owing to its toxins, mayapple was used medicinally by Native Americans and the settlers who learned from them. Extracts were used for cathartic, laxative, insecticide and suicide purposes.

Since the rhizomes of mayapple grow laterally and send up new shoots year by year, mayapples tend to grow in patches. Some colonies have as many as one thousand shoots. Usually all the individuals in such a patch are genetically identical. It has been found that in Princeton and some other localities, mayapples are not self-fertile. If pollination within a colony (done by human hand after the opening flower buds are bagged to exclude pollinating insects) results in fruit, there was more than a single clone present. In other areas, mayapples are able to fertilize themselves.

Mayapples belong to the barberry family. The genus *Podophyllum* (foot-leaf) is represented in North America by only this one species—*peltatum* (shield-like). It has many other common names: wild jalap, Indian apple, devil's apple, ground lemon, hog apple, raccoon berry and mandrake.

If beauty, potential food and medicine are not enough to stir up your interest in this plant, think of the legend associated with the European mandrake, an entirely different plant, belonging to the nightshade family. The roots of mandrake were trimmed to resemble a human figure and much used in medicine and magic. Mandrake was said to shriek when torn from the earth. Lest ill fortune befall the person who yanked out the root, its extraction was done by means of a dog and string. Early settlers transferred the name and magic to our native mayapple.

Blue Cohosh

Last April a Catskill mountainside glowed purple with blue cohosh (*Caulophyllum thalictroides*) shoots bursting out of the soil. Most likely, a few days later, clusters of five to twelve greenish-yellow-purple six-pointed flowers appeared, while the leaves were still dark purple and unfolding. I have records of blooming on April 23 in the Helderbergs and May 11 on Tongue Mt., but no data from farther north.

The purple color of the immature plant tends to overshadow the flower. Looked at closely, the half-inch flower has a geometric beauty with six sepals, six small petals and six stamens, all neatly aligned. The single short pistil matures before the stamens, which discharge their pollen through pores. Nectar produced by fleshy organs on the petals attracts bees that help cross-pollinate this plant.

After bloom comes the greening of blue cohosh as the plant grows to its full height of possibly three feet. Often the stem has a waxy bloom. If you rub it off, will the plant replace it? If you have a blue cohosh neighbor, try the experiment.

The plant bears one, two or three leaves, depending on which reference I consult. One author prefers to call the upper leaf, with nine to twelve leaflets, a bract. Thus one person's "two leaves" could equal another

Blue Cohosh

person's "one leaf plus bract." As usual, when I am writing essays in winter I am full of questions only the plants can answer.

The lower leaf, or leaves, is divided into twenty-seven or so leaflets. At the apex of each one- to three-inch leaflet are three to five pointed lobes. You'll notice a resemblance to the leaves of columbine and meadow rue, although larger size and a remaining bluish cast differentiate blue cohosh. The scientific name, *Caulophyllum* (from *kaulos*, stem and *phyllon*, leaf) *thalictroides*, comes from the resemblance to meadow rue, whose scientific name is *Thalictrum*.

Later in the growing season there is another bursting. The growing seed splits open the fruit that surrounds it and continues its maturation into a large blue "berry." It is actually a naked seed. Originally there were two ovules in the ovary of the pistil. One often aborts, but not always. Will close inspection tell us whether one or two seeds were involved in making the "fruit"? We'll have plenty of time for this observation, as the seeds hang on all summer and often persist after the leaves have fallen.

I will always think of Werner Baum, active ADK member, hiker, botanist and teacher, whenever I see blue cohosh, because it was he who explained this strange naked seed to me. (As far as I know all our other wildflowers ripen their seeds inside the ovary, which becomes a fruit.)

During winter, blue cohosh lives on in many cool, rich, moist forests as a thick rhizome under the ground, with the roots and scars of previous years' growth. Many Native American tribes used blue cohosh for medicinal purposes, particularly for easing childbirth and curing "female troubles." *Cohosh* is an Algonquin name; the settlers learned about its supposed value and called it squawroot and papooseroot. Medicine was made by boiling the root, or rhizome, in water. Not only are there potent alkaloids and glycosides in the rhizome, but the fleshy seeds are also listed as poisonous. Roasted, they have been used as a coffee substitute. The leaves may cause dermatitis.

Blue cohosh shares some characteristics with mayapple, and it should—both belong to the barberry family. Each is the only representative of its genus.

Violets

There are approximately fifty species of violets (*Viola*) in the northeastern United States. Many of these will greet you in spring in the Adirondacks with white, yellow and violet-blue flowers. You probably have violets in your own yard. Pick a flower. You will not harm the reproductive possibilities of the plant.

Examine the flower with a hand lens. It is bilaterally symmetric. Starting from the outside, there are five green sepals. Next there are five colorful petals. If whimsy appeals to you, "undress" the flower slowly (tweezers will make it easy) and you will expose a little person in a cape. This little person consists of five stamens closely appressed to the pistil. The conniving anthers make the cape. The stigma, the pollen-receiving end of the pistil, marks the little person's head. Note that s/he is having a foot bath: two stamens are equipped with appendages that secrete nectar into the spur of the largest petal. Now remove the stamens and examine the pistil. It has a crook in its style. This characteristic is shared by all members of the violet family.

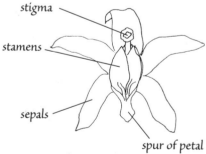

stigma

stamens

sepals

spur of petal

"Little Person"

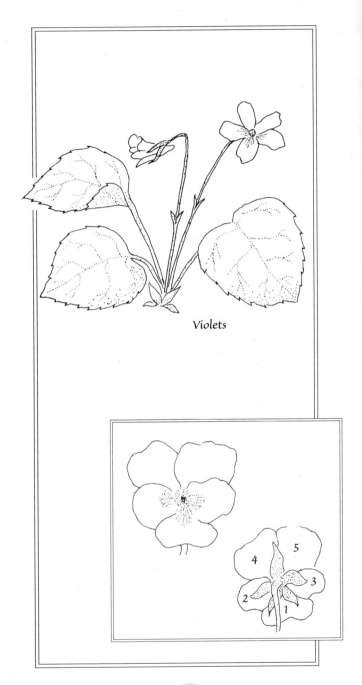

Violets

The violet flower with its beautiful insect guides, its supply of nectar, its attractive colors and, in some cases, its perfume invites insects to pollinate it. Take the time to linger and watch the insect's mode of operation. After pollination, the flower often matures into a three-sided fruit that bursts into three canoe-shaped segments when mature. The seeds are catapulted outward in most species.

fruit

pistil

Most violet species also produce cleistogamous flowers, hidden flowers that look like buds and never open. Charles Darwin first described them in 1877. Look for them after the blooming season. Self-fertilization occurs within the cleistogamous flower and fruits are produced. In a greenhouse one can experimentally induce open or cleistogamous flowers by adjusting the number of daylight hours.

If you grow pansies—big violets—no hand lens is needed for an anatomic examination. In *Reading the Landscape of Europe*, May Theilgaard Watts tells a German fairy tale about the flower based on its anatomy. Pansies are called *Stiefmuetterchen* (little stepmother) in German. In the old days stepmothers were reputed to be selfish, preferring their own children to their stepchildren. This stepmother's household was equipped with five chairs (sepals). Stepmother (petal 1) was fat and sat on two. Her daughters (petals 2 and 3) each grabbed a chair. The two stepdaughters had to crowd together on the remaining chair (see drawing, p 58). Of course there was a fairy who rewarded the good and punished the bad. The stepmother was given a beard and a hump on her back (the nectary); her daughters became bearded (the petals have hairs), but the virtuous stepdaughters were left velvety, without blemish.

Are you interested in eating wild plants? All parts of the violet are edible. We have a wealth of *Viola papilionacea*, the common blue violet, in our lawn and eat flowers, stems and leaves. They make edible decorations for tossed salads, cottage cheese, cakes and gelatins. The season for leaves extends into the summer. If you harvest flowers, you will encourage blooming for a longer season.

Collecting leaves can be destructive to plants. Over-harvesting can result in starvation because it is the leaves that carry on photosynthesis, making carbohydrates. Take only a few leaves from the center of each cluster. Euell Gibbons in *Stalking the Healthful Herbs* calls the violet "nature's vitamin pill" because it is high in vitamins A and C. Consult his book if you want to make violet jam or syrup or try some other concoctions.

You might like to try making a pH-sensitive solution from blue violets. Test it with an acid such as lemon juice. It will turn red. Try other tests and be surprised.

Even if you don't eat violets and don't entertain children with violet tales, you can enjoy the fleeting beauty of the flowers and the longer-lasting beauty of the leaves. Notice particularly how the small round leaves of the yellow violet, *Viola rotundifolia*, become the large round leaves that carpet some Adirondack woods in summer.

Three Trilliums

A forest in bloom with trilliums is a breathtaking sight. The genus *Trillium* was named for its triplets: three leaves, three persistent sepals, three petals (withering after two weeks or so of bloom), a three-celled ovary with three stigmas and twice times three stamens.

Occasionally there are mistakes. At Wolf Hollow, near Schenectady, my class found a large four-leaved white trillium and a three-leaved one with multiple petals. Petals had replaced stamens and pistil, making the flower look like a carnation. Though a member of the lily family, whose leaves generally have parallel veins, trilliums are net-veined.

Purple trilliums have maroon petals (occasionally also pink, salmon, green-yellow or white). The sepals are nearly the same size. Their alternate placement gives the effect of a six-sided star. This flower has a bad smell, and thus attracts pollinating carrion flies and beetles. Early herbalists, believing in the doctrine of signatures, used it for treating gangrene. I suspect it stinks only at certain times, as

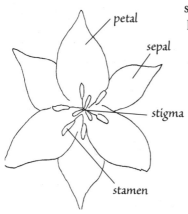

petal

sepal

stigma

stamen

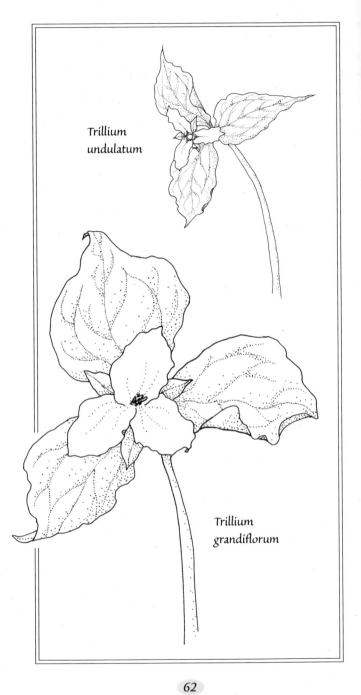

Trillium undulatum

Trillium grandiflorum

testing flowers has often had no unpleasant results. The fruit is an oval, reddish, ridged pod. The leaves are unstalked.

White trillium is our largest species, with white, wavy-edged petals that flush pink with age. It is pollinated by bees and butterflies. The sepals are narrower and shorter than the petals. The fruit is a red (some say black) pod with six angles or wings. The leaves are unstalked.

Painted trillium has white, wavy-edged petals with inverted pink V's at each base. It bears shiny, smooth red pods with three obscure angles. The leaves are stalked.

Purple trilliums (*Trillium erectum*) grow in the Adirondacks to an altitude of about 3500 feet. Painted trillium (*Trillium undulatum*) has been found up to 4400 feet. The former grow more often under hardwoods in what is commonly (and I think misleadingly) referred to as rich soil, while the latter grow more often under conifers in moist woods. In my home area, between the Catskills and the Adirondacks, purple and white trilliums vie with each other. White trilliums (*Trillium grandiflorum*) prefer basic or neutral soil.

Trillium erectum

Though there are some other species of trilliums in our area, the southern Appalachians have a much greater variety. All are perennials with thick, fleshy rootstocks. It takes a minimum of six years for a seedling to grow to reproductive age. Young trillium leaves are sometimes confused with young jack-in-the-pulpit leaves. Note the

A jack-in-the-pulpit leaf

difference in leaf arrangement and vein pattern.

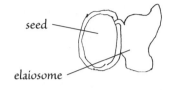

seed

elaiosome

It is interesting to follow the life history of a trillium plant as its flower matures to fruit. As the seed capsule ripens, its stalk bends to the ground. The seeds emerge. Each seed has an interesting fleshy outgrowth. It is a tissue, called an elaiosome or aril, adapted to dispersal by ants. Probably the attractant is a fatty acid—ricinolic acid. The cells are filled with oil drops and starch grains. Ants gnaw the elaiosome, but the rest of the seed is unharmed and can germinate away from the competition of the parent.

All this I read in Theodore Thomas Kozlowski's *Seed Biology*. A skeptical friend tested our book-found knowledge. She spread several of the crested trillium seeds on her lawn. Ants soon emerged from nest entrances. She watched one grab a seed and move it twenty-two inches in just over four minutes before carrying it into the dark underworld. Soon all of the seeds were gone.

Baneberry

Baneberry (*Actaea*) is an eye-catching forest perennial that is often difficult to identify because the flower lover has failed to notice it in modest bloom (when field guide identification is easier) and it now attracts attention with showy fruit. If you have trouble deciding whether the leaves are opposite, alternate, simple or divided you can find baneberries in the Nature Study Guild's *Berry Finder* and in the *Audubon Society Field Guide to American Wildflowers*, which features the berries. A skull warns you that this is a poisonous plant.

Native Americans warned the colonists about baneberries, and the various common names echo the warning; however, an infusion of the leaves was drunk by nursing squaws to stimulate the flow of milk. Among the Cheyenne, red baneberry was the highly esteemed "sweet medicine" (the name of a prophet-leader) and the root (underground stem) was steeped in water. I learned that from Jeff Hart's book *Montana: Native Plants and Early Peoples*. "Cheyenne believe that their children will grow up to be of good mentality, strong, patient—everything that has been attributed to Sweet Medicine's standards. Because the children are now drinking cow's milk, they are losing these qualities and becoming like cows." Sweet Medicine reputedly lived 445 years with the Cheyenne. Upon his death he transferred his sacred powers to this plant.

When checking on plants with a poisonous reputation I consult John Kingsbury's *Poisonous Plants of the*

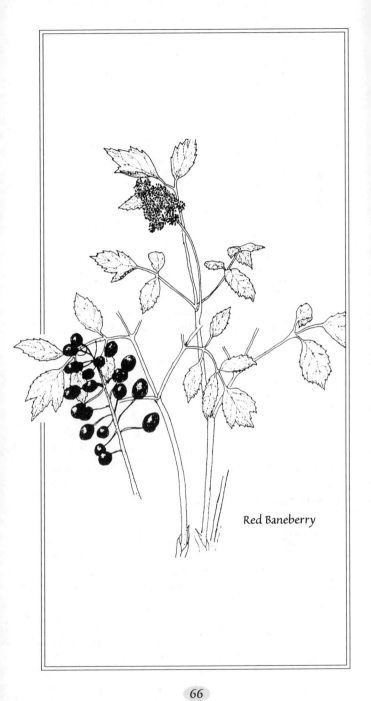

Red Baneberry

United States and Canada. He writes, "Members of this genus, particularly the European baneberry, *Actaea spicata*, are considered poisonous and references may be found in European works concerning the death of children after eating the conspicuous berries. Cases of loss of life, either human or of livestock, have not been recorded in the U.S."

Baneberries are one to three feet tall with leaves divided two or three times into sharply toothed leaflets. Flowers are congested on a short stalk with short pedicels (flower stalks). Both the stalks and the axis elongate later. The individual flowers are small and white. The calyx consists of four or five tiny sepals that fall as the flower opens. The corolla consists of four to ten small flat petals that soon drop. Numerous slender stamens remain. They produce the pollen that attracts bees, which pollinate the flowers.

In the center of the flower is the broad two-lobed stigma sitting right on the ovary. The stigma remains conspicuous as the pupil of a doll's eye on the fruit of white baneberry (*Actaea pachypoda*). Sometimes the dark purple color seems to leak out of the stigma onto the fruit.

The red baneberry (*Actaea rubra*) in our yard had twenty-six fruits. One dissected fruit had twelve one-eighth-inch-long, crescent-shaped dark-brown polished seeds (not rough ones as my reference book stated.) I forgot to record the number of fruits on my white baneberry. I dissected fruits of both baneberries on the same August day. The white one's seeds were greenish, unripe and

white
baneberry
fruit

only six in number, looking like tangerine sections.

Orra Phelps, ADK's long-time first naturalist, told us that she could tell the two baneberries apart by the veinier leaf of the red baneberry. Try that. Red baneberry begins to bloom about two weeks before white baneberry, but the flowering seasons overlap.

Linnaeus, the Swedish botanist whose ambition was to give all living things a genus and species name and fit them into his classification scheme, described an American variety of *Actaea spicata* with white berries. No wonder there has been some confusion. Variation of color within a species is common. The two species may also cross. Last September, among Canada violets in late bloom and ginseng and spikenard in fruit, I saw a red-berried fat-red-stalked baneberry. Was it a hybrid?

Indian
Cucumber Root

Indian cucumber root is a woodland plant that gets its common name from its edible rhizome (the horizontal, underground wintering stem), which does taste like cucumber and was once used for food by Native Americans. Because we now have more humans and less forest and the plant must be dug up and destroyed to serve as food, we should regard its edibility as interesting history and enjoy the living plant. I was curious about the edibility of the fruits. James Walker Hardin and J. M. Arena in *Human Poisoning from Native and Cultivated Plants* say, "unknown, caution!"

The scientific name of Indian cucumber root is *Medeola virginiana*. The generic name apparently comes from the Greek mythical sorceress Medea. Perhaps Linnaeus, who gave the name, heard wondrous medicinal tales from his collectors and informants in America. Our species seems to be the only *Medeola* in existence and it is confined to eastern North America.

You will find Indian cucumber root in two forms, one-storied and two-storied.

rhizome, 3"

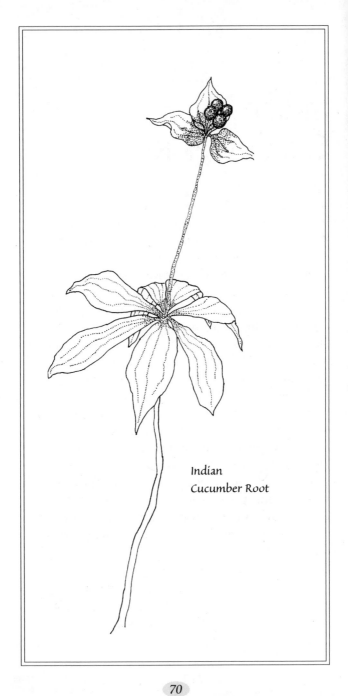

Indian
Cucumber Root

All my reference books are silent about the plants with one whorl of leaves! Some folks confuse this immature one-story form with starflower (*Trientalis borealis*, in the primrose family). The size and habitat can be the same. Both have a whorl of five to nine leaves, but note the difference in regularity of leaf shape and placement and especially the difference in venation. The parallel leaf veins of Indian cucumber root are

starflower

characteristic of the lily family, of which it is a member. Flowers in groups of three to nine grow from the axis of the upper leaf whorl on one-inch stems. They usually droop. They, too, follow the lily prescription:

three sepals

three petals (often sepals and petals look similar)

six stamens

one pistil, with three styles

In Indian cucumber root, the three conspicuous long styles carry the pollen-receptive surface along

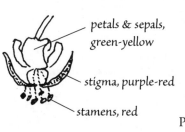

petals & sepals, green-yellow

stigma, purple-red

stamens, red

their inner side. After pollination the flower stems straighten and fruits ripen to purplish-black berries above the whorl of leaves. Purple stains often

appear on these leaves, and sometimes on the stem and lower leaves also, as if spilling from the fruit. I have noticed the stain appearing while the berries were still green. While keeping in mind that everything need not be adaptive, it is fun to speculate about the adaptive advantages of the characteristics we notice. Could the dark stain help in heat absorption and hasten the ripening of fruit? Could it help to attract birds who could distribute the seeds?

Last September, along the wide path to Lost Pond in the Pharaoh Wilderness Area, we saw a beautiful display of fruitful Indian cucumber root, many plants bearing six or seven berries. Ferns were tan and brown, as were the lily cousins— bellwort, Solomon's seal and false Solomon's seal. The eyecatchers were hobblebushes with their purpling leaves and Indian cucumber root. Back from the trail there were fewer fruiting plants. I suspect many forest plants respond to disturbances that admit more sunlight by increased sexual reproduction.

As I write this I am full of questions that cannot be answered now, but perhaps you will enjoy doing some investigating. Does the flower have an odor? What insects are attracted to pollinate it? Can it self-polli-nate? Does it propagate asexually from the rhizome? Are the colonies clones? What is the proportion of immature to flowering plants? Can a seed produce a flowering plant in one season? Does a stout rhizome always produce a two-storied plant? I would enjoy hearing from readers who have found answers.

Lady's Slippers

From mid-May into early June, attentive Adirondack hikers will find nature reduces their walking speed to a stuttering pace. Photographers will find the most frequent progress stoppers will be masses of pink lady's slippers (*Cypripedium acaule*). They grace the trail from the Deer Leap trailhead on NY 9N south along the ridge of Tongue Mountain. They clump in the dry lichens beneath the evergreens on Crane Mountain and on Round Mountain. These orchids are spectacular and interesting natives and they can be found beside many Adirondack trails. They are most abundant on acid sites where conifers outnumber hardwoods. Although not really rare, they are protected in New York. Your observations will not offend them.

Orchids evolved from lily-like plants. Flowers of the Lily family are six-parted, with three outer parts (sepals), three inner parts (petals), six stamens (carriers of pollen) and one pistil (which ripens into the fruit). Each part is separate from the others. In the highly developed orchid, parts have become fused and specialized.

The following description of the lady's slipper flower is borrowed and paraphrased from a 1905 book, *Our Native Orchids*, by William Hamilton Gibson. One sepal stands directly over the pouch like a banner; the other is composed of two fused sepals, directly under the pouch. Two of the petals are the wings that fly at right angles to the banner and the

Lady's Slippers

third petal is the pouch, or lip or slipper. Stamens and pistil have grown together into one organ called the column. A sterile stamen makes a triangular roof covering the opening to the lip so the two pollen-filled stamens and the comb-like tip of the pistil are kept dry. The pollen is a sticky mass. Gibson continues with:

> Once the bee is in the slipper and has sipped the secretion that exudes from the hairs that line the inner surface of the lip, he finds that those pink-veined edges that lured him in curve hopelessly away from daylight, but the hairs inside serve as guide. They all point back and up to a pair of narrow side doors—a plainly marked exit—through which he can squeeze. As he crawls past either open anther he will surely be smeared with pollen. However, he will be scraped clean by the pistil that hangs like an inverted doormat at the entrance to the next blossom. The object of the brilliant veins and spots around the incurved opening of the pouch is to coax the bee to squeeze under the pistil, thereby brushing it with the pollen of the last flower visited. So, he performs the rite of cross-fertilization.

Gibson's observations make me wonder if there are male pollen-collectors among our wild bees. If he watched honeybees, he mistook their sex. All workers are female. He probably watched bumblebee queens, as more recent observers have. They also have failed to find any food reward within the slipper. This may account for the low rate of pollination and fruit set. A disappointed bee is less likely to make repeat visits. Maybe you'd like to emulate Charles Darwin and, armed with a sharp pencil, a brush or a dead bee, imitate the action of an insect pollinator. Darwin's observations and experiments are described in his *Fertilization of Orchids*, a most enjoyable book.

You have now set the stage for the production of a capsule, filled with thousands of tiny seeds, a compensation for fruit scarcity. The capsule is an object of beauty. It becomes woody and can be enjoyed for a year or two after it has split and released the seeds. These seeds must find a fungal partner for germination and establishment in the wild. The fungus nourishes the embryo, which has no food reserves, until it becomes a photosynthetic plant.

The scientific name of our flower is *Cypripedium acaule*. *Cypripedium* comes from *Kypris*, a Greek name for Venus, and *pes* or *pedis* meaning foot. A venus slipper in Greece became a moccasin flower in America. It was a Christian custom to make over the personal property of Venus to the Virgin Mary, thus our name lady's slipper. *Acaule* means stemless, an apt description of the two basal leaves.

Orchids belong to an enormous and fascinating family of plants. Kudish (see Selected References) lists twelve species of orchids in the Adirondack area covered by his book, *Adirondack Upland Flora*. I have noticed more books on orchids in the library than on any other plant group.

Bunchberry

Bunchberry (*Cornus canadensis*) is one of our native Adirondack dogwoods that we can enjoy in all stages as we climb back into an earlier season all the way to timberline. A native of North America and eastern Asia, it likes acid soil. Although the plant is quite tolerant of shade, it flowers most abundantly in open areas. New shoots of this perennial rise from slender horizontal rhizomes; thus we find bunchberry in colonies.

The leaf- and flower-bearing stems are erect, three to nine inches high, and bear a cluster of four to six (sometimes more) apparently whorled leaves at the summit. Below the whorl are one or two pairs of smaller leaves or scales. The leaves are pointed at both ends. Two or three pairs of lateral veins arise from the center vein below its middle. Note how the veins curve gracefully to parallel the smooth leaf edge. This is characteristic of all dogwoods. Another shared feature is the presence of latex in the veins. A scientifically minded friend tested viburnums for latex and found some in at least one species. The test consists of gently pulling a leaf apart. I now doubt the diagnostic value of latex for "confirming" dogwoods.

The generic name of dogwood is *Cornus* (meaning *horn* and referring to the hardness of the wood of some dogwoods). All dogwoods are trees, shrubs or herbs with opposite leaves except one species, a common inhabitant of our woods, *Cornus alternifolia*. Its name, alternate-leaved dogwood, testifies to its

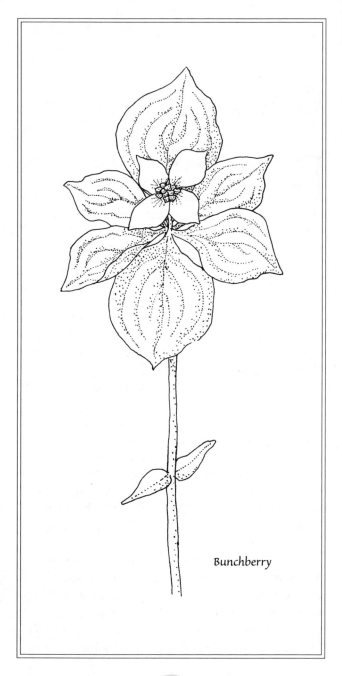

Bunchberry

uniqueness. I have torn a leaf from all our woody *Cornus* species—red osier, silky, alternate-leaved, gray-stemmed, round-leaved and flowering—to demonstrate the elastic white thread (the latex) in their veins, but I cannot recall testing the leaves of *Cornus canadensis*, our bunchberry. Here is an easy research project for you. You need tear only the tip of one leaf.

Using a hand lens, you can see that the flowers of bunchberry have four sepals, four petals, four stamens and one pistil. They grow in a dense globose cluster at the tip of the plant. The cluster is surrounded by four showy white bracts. If you climb a mountain or begin your bunchberry watch early enough you can see the transformation of four thick, purplish bud scales into these beautiful white bracts. While the bracts remain to advertise to pollinating insects, some of the true flowers within are transformed from buds to open flowers, pollinated flowers and young fruit.

Palmer and Fowler, in their *Fieldbook of Natural History* (see Selected References), illustrate a fruited plant that has four leaves in a whorl beneath the cluster of fruits. In my experience there have been four leaves on sterile stems, six or more on fruiting stems. I will look more critically this summer and remind myself that a look at the plant is more truthful than a look at a book.

The fruits are red globular drupes, each containing one smooth seed. Fruits and bracts of our two showy-bracted dogwoods, flowering dogwood and bunchberry, closely resemble each other. So I will quote Edmund Stiles: "The fruits of flowering dogwood which mature in September and October are rich in fats. Up to 35 percent of dry weight of pulp may consist of lipids. Such nutrient sources are much sought after by migrating birds. The high lipid level also makes this fruit attractive to microbes. If not eaten by birds the fruit will probably rot in place, the seeds undisbursed." I find the fruits tasteless, but a feast for the eyes.

Joe Klimas, in *Wildflowers of Eastern America*, writes that because the berries seemed to stimulate the appetite, Scottish immigrants called bunchberry "plant of gluttony." If bunchberry fruits had a reputation as appetite depressants, think how popular they would be today. We have few alternate names for bunchberry. I know of only dwarf dogwood and dwarf cornel. (Clem Habetler, the illustrator, found more: crackerberry, puddingberry, rougets.)

I recently met a puzzling, exotic dogwood on a college campus. Its red fruits were compound and soft, like fat strawberries. *Cornus kousa*, like most Asian dogwoods, may have evolved from the simple-fruited American dogwood type.

Richard Eyede (now deceased) of the Smithsonian Institution thought that selection for big, compound fruits was the result of contact with Old World monkeys, like macaques, who can distinguish colors, prefer "more goodies per grab" and scatter the seeds. America retained the older, simple fruit because New World monkeys are blind to red.

Summer

Only a few woodland flowers bloom now; partridgeberry, wood sorrel, dalibarda, pyrola and Indian pipe come to mind. Now is the time to enjoy the variety of designs that have evolved in green—the leaves of flowering plants and the nonflowering ferns.

Then turn your attention to sunny places such as cultivated and abandoned fields, roadsides and wetlands. In addition to the flowers featured in the following essays, you will certainly meet clovers, bird's-foot trefoil, daisies, Queen Anne's lace, daylilies, chickory, butter-and-eggs and many other recent dryland immigrants. Natives such as milkweeds, fireweed, cattails and common elderberry bloom lustily in summer. Bogs will delight you with a variety of orchids and other special plants. Check riverbanks, lake shores, and marshes for plants especially adapted to those habitats.

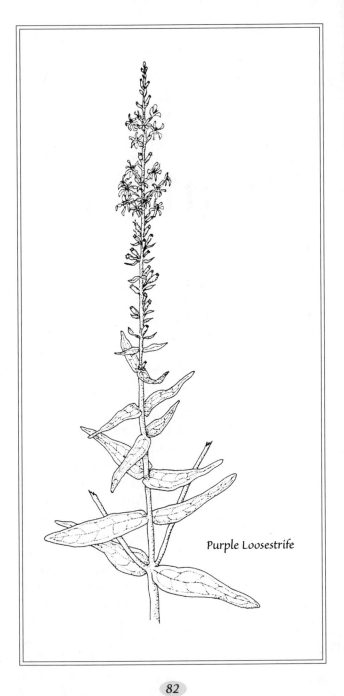

Purple Loosestrife

Pretty Perplexing Purple Problem: Loosestrife

B eginning in June, I get phone calls from people asking, "What is that beautiful purple flower along the highway, in ditches and around ponds?" The inquiries have lessened lately, as more people have seen pictures of loosestrife (*Lythrum salicaria*), along with dire warnings of its evil intentions. Headlines, such as "Subduing Purple Loosestrife" in the August 1994 *Conservationist*, alert people. The Nature Conservancy calls it "Deadly Beauty." *National Wildlife*, October 1982, headlined its purple loosestrife article, "Renegade Plant Squeezed Wildlife." Local newspaper headlines shout, "Harmful Flowering Plant Threatens to Overwhelm America's Wetlands." Thirteen states list it as a noxious weed. You are warned not to buy and plant the garden cultivar advertised as sterile. It might not be sterile enough. Stanley Smith, former curator of botany at the New York State Museum, had a more balanced view. His article was entitled, "Purple Loosestrife—Weed or Beauty." I am tempted to steal that title, changing "or" to "and."

The name *loosestrife* probably means reducing discord. It should soothe tempers. Indeed this plant was brought from Europe purposely, as well as

accidentally, early in the nineteenth century. It was
used as a medicinal herb for diarrhea, ulcers and other
ills. It flourished in America. It is a perennial, fast
growing (over one centimeter per day) and able to
reproduce vegetatively as well as by its prodigious seed
production. A young plant might produce ten thousand seeds, a mature plant two and a half million. The
seeds are long-lived and easily spread. Like many new
plant immigrants, purple loosestrife has no predators
in its new habitat.

To control very recently established purple loosestrife, careful hand pulling might do, if all parts are
removed. The plant can sprout even from cuttings of
the main stem. Among unsuccessful control methods
have been flooding, mowing, cutting, herbicides and
burning. To illustrate the problem: fire stresses native
plants more than purple loosestrife and gives loosestrife seeds that are stored in the soil a chance to
sprout. Weevils and beetles, selected from European
plant-eating insects for their host specificity, have been
introduced and are expected to consume 80 percent of
"the crop." *Scientific American* (September 1995, "One
Good Pest Deserves Another" by Sasha Nemecek)
quotes Richard A. Malecki of the National Biological
Survey saying, "There have been small, scattered
success stories since the insects were introduced in
1992." He did not foresee nationwide impact for
fifteen to twenty years. Importing more exotics to
control exotics is always risky. Eating habits may
change.

At the Natural History Conference held at the New
York State Museum in 1994, I learned from Spider
Barbour of *Hudsonia* that purple loosestrife has been
in the Hudson Valley for 160 years. Cecropia,
Polyphemus and Io moth larvae are now eating the
leaves of this alien plant. Cocoons have also been
found on purple loosestrife. Honeybees, bumblebees

and butterflies have been seen feeding on pollen and nectar. Purple loosestrife is the source of much appreciated honey. At a nearby beaver pond, in low water after a dry summer, we noticed deer had sloppily grazed both cattail and purple loosestrife. Maybe patience is the best approach to the problem of the purple plague.

For us amateur botanists, purple loosestrife is an ideal plant for observation and fearless experimentation. In winter, look for small reddish buds at the base of dead flower stalks. In spring, watch new plants emerge, looking like fat, scaly asparagus stalks. The stalks elongate and display three different kinds of leaf arrangements: pairs, whorls of three and spirals. You may need to check several plants to see this. Notice that the cross section of stems varies along with leaf arrangement. The leaves, round or heart-shaped at their base, are usually hairy. There is much variation. The plant can grow to be one to seven feet tall. But most fascinating are the variations among the four- to six-parted flowers. If you search for a while you should be able to find three different types (in three different plants, because those on one plant should all be identical):

> Long pistil, with short and medium stamens
> Medium pistil, with short and long stamens
> Short pistil, with medium and long stamens
> Long stamens bear brilliant green pollen, while
> medium and short stamens have ordinary
> yellow pollen.

Darwin studied this flower, experimented with it and described it beautifully in his book, *Fertilization of Flowers*. With careful measurements Darwin found that pollen grain size varied directly with stamen height, the largest pollen grains being the green ones. He weighed seeds too and found that five long-pistil seeds equaled six mid-pistil seeds or seven short-pistil seeds. Darwin found the three types of plants in equal numbers. When protected from insects with netting, only a few seeds were set by each type. This showed the need for cross-pollination and the fact that no rule is absolute. Playing Cupid with a camel's hair brush, Darwin (always giving credit to his children and whoever else assisted) made eighteen distinct pollinations and showed that the union of sex cells from pistil and stamen of equal length are alone fully fertile and "legitimate." Darwin gives vivid descriptions of how insects are dusted in different parts of their anatomy by pollen from different length stamens. All this and more you can easily read, though it was written in 1877. Enjoy your own explorations with this gorgeous, common plant and Darwin's conclusion: "In short, nature has ordained a most complex marriage-arrangement, namely a triple union between three hermaphrodites."

A Native Weed: Evening Primrose

E vening primrose (*Oenothera biennis*) is a native belonging to an aggressive group of plants that exploit new landscapes and are known to us as weeds. It is widespread from Canada to Florida and westward, and is common along roadsides and in old fields. Last summer I noticed an abundance near Saranac Lake. Michael Kudish of Paul Smith's College speaks of a sweet lemon scent, but I've missed that.

In its first season the plant forms a rosette of leaves, effectively cutting off sunlight to nearby competing seedlings. These leaves are four to eight inches long, often with a reddish cast and decorative veins. They are attached to the top of a stout branched root.

After a restful winter under the snow, this biennial resumes activity in the spring of its second season. A stalk grows from the center of the rosette, bearing leaves with clasping bases. It may grow to six feet. It branches near the top, where it bears light yellow flowers. The flowers open between four and ten in the evening. The opening of the flower is well worth watching. The petals seem to struggle from the tightly embracing sepals. After the petals escape, the sepals bend backward and fold against the flower tube. Getting at the nectar in this tube requires a long tongue, such as that of a butterfly or moth. Below the long flower tube is the ribbed ovary. If the flower is pollinated during its one to two days of bloom, the

Evening Primrose

ovary may develop into an erect, cylindric, ribbed seed pod.

The evening primrose family has an interesting flower plan based on four, rather than the more common five or three. There are four sepals, four petals, eight stamens and a pistil composed of four fused parts. The stigma opens into a cross and usually hangs out at one side of the eight anthers.

Once, while enjoying evening primroses, I noticed a pinkish-yellow moth within a pink-tinged fading flower. What a wonderful example of protective coloring! Trying to find the identity and ecological niche of this creature I turned to an old favorite, *Handbook of Nature Study* by Anna Comstock. Here she tells of a green caterpillar, looking like the bud of the evening primrose, nibbling on it, damaging the flower within. Eventually the caterpillar descends to the ground, burrows and overwinters in the soil as a pupa. Next summer emerges a beautiful moth, *Alaria florida*, the moth I saw. "This moth is the special pollen carrier of the evening primrose; it flies about during the evening and thrusts its long tubular mouth into the flower to suck the nectar, meanwhile gathering strings of pollen upon the front of its body. During the day it hides within the partially closed flower, thus carrying the pollen to the ripened stigmas, its colors meanwhile protecting it almost completely from observation."

Native Americans ate the roots of the evening primrose and dried them for winter use. It was introduced to Europe in the seventeenth century and cultivated in England as a root vegetable. For cooking suggestions, consult Alan Hall's *Wild Food Trailguide* (see Selected References).

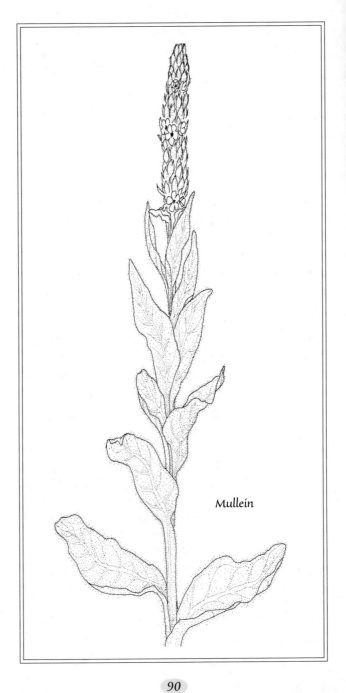

Mullein

Mullein

Toward the end of summer, wildflowers in bloom become scarce in the forest. It is now that roadsides, railroad right-of-ways and other disturbed places come into their glory. One striking denizen of such dry areas is mullein (*Verbascum thapsus*), an immigrant from Europe, where it is a garden plant. Here we regard it as a weed. You can find it naturalized from Nova Scotia to Florida to California. It is a biennial, producing a leaf rosette the first year, and flowers the next. Flowering plants can grow to eight feet with alternate, densely wooly leaves that can be over a foot long. Their stems extend down the stalk as flutings.

I am writing this the first week of July. Mullein has just begun to bloom. It will continue to bloom through September. The flower stalk has a scruffy appearance. The expanding tip is still all buds, but below there is a flower here and there interspersed with buds (flowers to be) and developing fruits (flowers that were). If there is a pattern, I haven't found it.

Look at a flower with a magnifying glass. You will find a five-

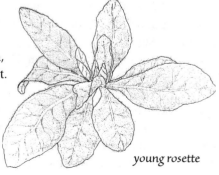

young rosette

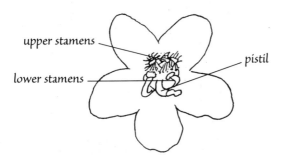

upper stamens
lower stamens
pistil

lobed calyx buried in the spike, five fused yellow petals
with lobes arranged in bilateral symmetry, one green
pistil in the center and five peculiar stamens around it.
The three upper ones are short and hairy. After they
have shed their orange pollen, the two longer,
smoother, lower stamens curve upward and present
their pollen to insect visitors. The flowers can also self-
pollinate. The fruit is a capsule. When it is ripe it
opens partially and shakes out many tiny seeds, each
looking like a section of a corn cob. I brought a
twenty-seven-inch top section of the flower spike into
the house and popped it into a container of water
which promptly fell over. The section weighed half a
pound. I gave it a shower, curious to see whether the
felty leaves would absorb water. They shed it. The
outer layer of the stem peeled off easily with the
flutings. Beneath was a ring of vascular tissue—the
internal transportation system of the plant. Within
that was a solid pith, good for cutting into shapes and
printing designs.

The name mullein comes from the Middle English
moleyne, which means soft. Common names include
flannel plant, velvet plant, feltwort, Adam's flannel,
blanket leaf. Take a good hand lens or use a micro-
scope to examine its tree-like hairs. Some people get
dermatitis from contact with the hairs. I experimented
and put two leaves in my sandals. After an hour my

feet burned. A mild case of dermatitis develops on the cheek after rubbing with a mullein leaf, the source of another name—Quaker rouge. The name hag's taper derives from an old use as night lights. Hairs from the leaves were twisted into wicks, inserted into mutton fat and lighted. The names torches and candle plant come from the time when the whole spike was dipped into fat and burned.

My fragment of mullein has not a single hole—no animal bit through the protective hairs—but it is full of live and dead insects and their droppings. Do the inhabitants of mullein go elsewhere for meals?

Mullein is well protected from drought by its hairy leaves and its deep tap root. Its winter rosette hugs the ground and survives under snow. It is a tough plant with which we can experiment as we ask it questions. Does anyone know whether the flowers bloom only one day and whether they "sleep" at night?

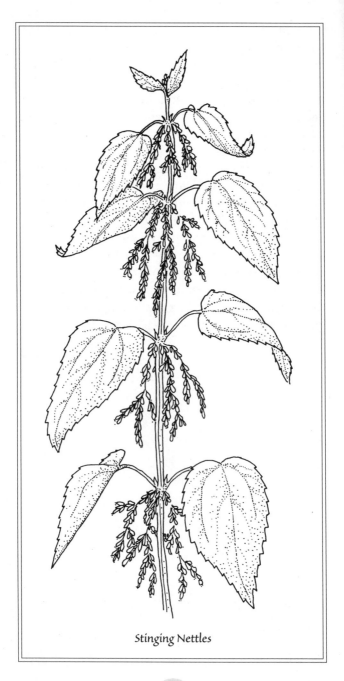

Stinging Nettles

Stinging Nettles

The first time I met nettles with alternate leaves was on the northern end of the Tongue Mountain trail, in a seepage area near the outhouse. It was late July, and the nettles were blooming.

None of my field guides allowed nettles to have anything but opposite leaves. In my encyclopedic book at home I discovered that nettles with stinging hairs and alternate leaves could be *Laportea canadensis*. If the leaves are opposite and the flowers are four-parted, the nettle belongs to the genus Urtica, from the Latin *uro*, to burn. *Urtica dioica* is naturalized from Europe and Asia, while the taller, slimmer and less stinging *Urtica gracilis* is a native. Some botanists consider *U. gracilis* to be a subspecies of *U. dioica*. All three are perennials.

If you have never had intimate contact with a nettle, you owe yourself the experience. Urtication, stinging the flesh with nettles, was the herbalists' cure for rheumatism. There is no delayed reaction, as with poison ivy, and therefore victims learn to identify the offending plant easily. The reaction wears off in twenty minutes or so. It seems hardly worth finding plantain, crushing it and squeezing its juice on the inflamed skin, as Euell Gibbons suggests in *Stalking the Healthful Herbs*.

The tapering tube of a stinging hair ends in a bent position with a narrow constriction whose walls are impregnated with silica. When you brush against it

stinging hairs

the tip breaks off and the fractured tube enters the skin and injects venom from the swollen base of the hair. One author advises rough handling, which causes the hair to break lower down and not penetrate the skin. Try it, or use leather gloves to harvest young nettle plants.

Gibbons waxes enthusiastic about the nettle's flavor as a potherb as well as its high vitamin A and C content. Cook the washed plants for twenty minutes in a covered pot in the water that clings to them. The stinging element is destroyed by cooking. Despite the advice of other culinary authorities, Gibbons does not approve of summer leaves from the top of the plant, as they contain gritty particles. He suggests that you save the cooking liquid and drink it or use it as a hair tonic!

A rennet substitute for solidifying milk can be extracted from nettles and a yellow dye can be made by boiling the plant. Livestock thrive on dried, mowed nettle.

Mature stems yield a strong fiber. In their *Fieldbook of Natural History*, Laurence Palmer and Seymour Fowler speak of the cultivation of *Urtica dioica* in Europe during World War I for making fabrics for tents and clothing. There is a Grimm fairy tale I remember from my European childhood in which a girl needs to make six shirts out of nettles to turn six swans back into her brothers. Checking on my memory in the local library, I found translations (six of them) charging the girl with making shirts of starflowers or asters. Are nettles X-rated?

May T. Watts, in *Reading the Landscape of Europe*, comments on the frequent nettling a botanist endures and the preference of nettles for nitrogenous soil: "In France [nettles occurred] at the base of every wall, but in England we felt the pricks only where the wall was somewhat secluded. Could the nettle distribution be indicating a difference in the relative sense of propriety with which Frenchmen and Englishmen relieve themselves? In Denmark, nettles are said to grow where innocent blood has been shed. In the Scottish highlands they are said to grow from the bodies of dead men." Eastern Europeans believed nettles to give protection from thunder and witches. Nettles have even been recommended as an aphrodisiac.

The *Urtica dioica* subspecies *gracilis* from upstate New York was the subject of a sex allocation study by Joanna M.K. Smith, a graduate student of the State University of New York College of Environmental Science and Forestry. Before giving you a quick review of her discoveries, let me explain that some plants, such as nettles, have separate male and female flowers. Nettles can be monoecious, with both kinds of flowers appearing on the same plants. The native stinging nettle is mostly monoecious, the Eurasian nettle mostly dioecious. Sex expression in plants has a strong environmental as well as genetic component. In most plants, female flowers—potential fruits—require a larger investment of resources than male flowers. We expect male-biased flower ratios in poorer habitats.

male

female

Smith found that weaker plants had a definite female bias; in fact, all plants shorter than 100 centimeters were female, while most of those taller than 125 centimeters were monoecious. This suggests that males are costlier to produce. Sure enough, the dry weight of a male flower is more than twice that of a female flower in fruit. You may wish to do your own investigating with a hand lens and gloves.

Nettles also reproduce asexually by underground stems, forming large clones. They are genetically identical and good material for studies involving the effects of changes in nutritional status.

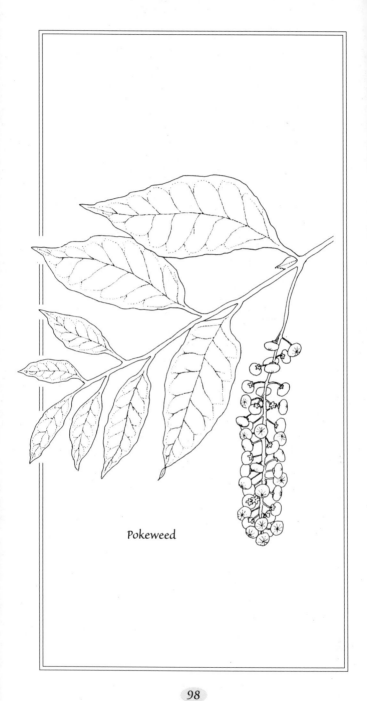

Pokeweed

Pokeweed

Pokeweed is a tall, coarse plant with leaves up to twelve inches long and smooth reddish stems. It grows on roadsides, in cultivated fields and waste places from Minnesota to Maine and south to Mexico. Amazingly, given its habitat, pokeweed (*Phytolacca americana*), also known as poke, skoke and inkberry, is a native. Flowers begin to appear in June but you can often find them later, along with the fruit. I watched a raceme (an axis with stalked flowers along it) begin to bloom when it was two and one-half inches long. It doubled in length by the time the flowers at the tip came into bloom.

The calyx is white with five rounded sepals. There are no petals. There are five to thirty stamens and a ring of five to fifteen pistils. Wild bees and flies pollinate the flowers. Most flowers can also self-pollinate. Each pistil has one style and stigma. The ovaries swell so much as they mature to a berry that they engulf most of the styles, which then look like little whiskers. The berries and their stems are an electric pink-purple.

Dye has been made from the berry juice. When I wrote with it, the ink was not permanent. One book tells me that the Portuguese colored port wine with pokeweed berries from plants imported from America. "Port wine" has also been made from whiskey and pokeweed berries. While we are on the subject of alcoholic beverages—it has been noticed that birds can become intoxicated on the berries. Drunk or not, they spread the seeds. Fruits become tan and reduced in size toward the end of the growing season if they contain aborted ovules. The abortions are blamed on pollen grains infected with fungus.

The stem interior fascinates me. It consists of a stack of discs. Similar energy-saving sturdy construction is found in the twigs and stems of butternut and walnut. The roots of this perennial can grow to six inches in diameter. Both roots and berries are considered quite poisonous for humans, acting as a slow emetic. That means vomiting occurs about two hours after eating. Yet you will find pokeweed listed in most books on edible plants. *A Field Guide to the Edible Wild Plants* by Lee Peterson warns you to use only shoots that are less than six inches tall and to discard or peel those tinged with red, then boil twenty to thirty minutes in at least two changes of water. You then have a "wild aspara-gus." A Tennessee journal says commercially prepared "poke salet" cans are available in local groceries. I am waiting for a can to come my way.

Asiatic and African species of *Phytolacca* have been investigated and used medicinally for a long time. In the fall of 1990 our newspapers and journals carried articles about pokeweed with messages such as, "Drug developed from poisonous weed appears to be a thousand times more potent than AZT in destroying the AIDS virus."

For once I was richly rewarded while looking through botanical journals for recent research reports about my plant. In one report about seed germination, juice from pokeweed fruits completely inhibited germination; root extracts were less inhibiting and extracts of juvenile leaves were least inhibiting. When birds consume the seeds they very nicely remove the inhibition by eating it and then they eliminate the now uninhibited seed at some distance from possible parental root inhibition.

If you are curious about the word *poke*, it does not seem to refer to the vigorous way the shoot appears, but comes from a Native American name for blood— *pokan*. Pokeweed is a potent and promising plant.

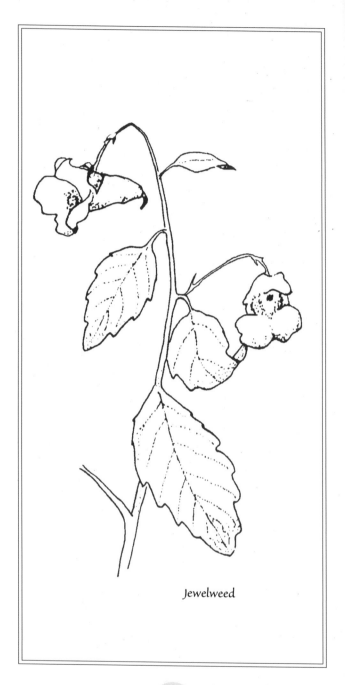

Jewelweed

Jewelweed

So many of our native wildflowers are perennial plants that an annual New York plant stands out: jewelweed. In wet areas you can find two species of jewelweed: the spotted one with orange flowers (*Impatiens capensis*) and the pale one with yellow flowers (*Impatiens pallida*). The leaves of both are covered with fine water-resistant hairs (use a magnifying glass to observe them).

Dunking the leaves in water or observing them after rain or dew may make you decide that the name jewelweed comes from the beauty of water beads on the leaves. Plant stems are translucent and hollow except at the joints. No wonder they are hollow: during one short season this plant can grow up to six feet. The root system is shallow. None of my books mentions the auxiliary support system, but I have seen many jewelweeds with prop roots just like the ones corn grows. You can watch these organs develop on the lower stem as red pimples that stretch out to "noses" and lengthen rapidly to roots.

The flower has three sepals, one of which is equipped with a horn-shaped nectary, and three petals. Above the entrance to the nectar sac, there is a knob

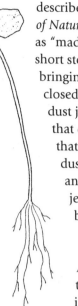

described by Anna Comstock (*Handbook of Nature Study*, see Selected References) as "made up of five chubby anthers; their short stout filaments are crooked, bringing the anthers together like five closed fingers holding a fistful of pollen dust just ready to sift it on the first one that chances to pass below. Thus it is that the bumblebee gets its back well dusted with the creamy white pollen and does a great business for the jewelweed in transferring it." The bumblebee transfers the pollen to another flower, one that is older, in the next stage of development. After releasing pollen for about twenty-four hours, a flower sheds its anthers and presents its pistil, ripe and ready to receive pollen for about four hours. This tandem ripening of sex organs is considered to be one of the many mechanisms by which flowers discourage self-pollination. Hawkmoths and hummingbirds have also been observed as pollinators of jewelweed.

The flowering season lasts a long time, so one can find flower buds, flowers and fruits at the same time. Fat fruit pods explode at the slightest touch and split lengthwise into five ribbons, expelling the seeds. This explains the generic name Impatiens and the common name touch-me-not.

The seeds, when peeled, are robin's egg blue. Maybe this is the source of "jewel" in jewelweed! The seeds will lie dormant through the winter and germinate in April and May, making a bright green carpet of twin leaves. These first leaves are the cotyledons, or seed leaves. Between them soon appears a stem with true leaves arranged in pairs. After two to four pairs of leaves have developed and the seed leaves have dropped, the stem begins a zigzag course of growth. At each bend one leaf appears. Thus, jewelweed switches from a pattern of juvenile opposite leaves to adult alternate leaves. By June the lower leaves bear leafy branches in the axils, and the upper leaves bear branches with leaves and flower buds. Soon flowering occurs, and the annual life cycle repeats itself.

It was through Orra Phelps's example that I began to eat and enjoy the ripe seeds of jewelweed. I think they taste like butternuts. Try them. Many of these seeds are the result of self-fertilization taking place inside buds that never open. Like violets, jewelweeds have cleistogamous (hidden) flowers. These flowers are tiny, only one to two millimeters across. They grow singly from the axils of leaves on lower branches and have no nectar and only a few pollen grains. The capsules contain one to three seeds rather than the three to five seeds of the open-flowered fruits. The fruits take only three weeks to mature instead of five.

The advantages of cleistogamy as a survival strategy have been beautifully documented. As an annual with short-lived seeds, jewelweed is absolutely dependent on efficient pollination to propagate itself. In ideal growing situations—moist and sufficiently sunny— jewelweed grows lushly, producing many showy flowers, offering rich sugary rewards to pollinators. In drier, shaded locations jewelweed remains stunted, produces only cleistogamous flowers and soon dies, but it is often able to maintain itself in such an environment year after year. The cost, in energy, of producing a seed through cleistogamy is only one-third to two-thirds of the cost of seed production through the showy flower method. It has been noticed that, even in favorable habitats, grazing or pruning will cause the plant to produce only cleistogamous flowers.

If you have a patch handy, you could experiment with jewelweed's emergency responses. Self-fertility is, of course, a great advantage for the single pioneer seedling in a new site. On the other hand, outcrossing, the expensive reproductive strategy, generates new genetic combinations and plays a role in adaptation to an ever-changing environment. Jewelweed has taken both roads to success.

Euell Gibbons and other authors suggest eating young jewelweed stalks as potherbs. The use of jewelweed as a preventative or cure for poison ivy rash is still controversial. One person I know purposely exposed the backs of both hands to poison ivy, then rubbed a handful of mashed jewelweed over one hand. A week later I received a card: "Neither hand developed a rash. Back to the drawing board."

Jewelweed is host to various guests: insect galls on leaves and flowers, stalk borers inside the stem, leaf miners between the upper and lower surface of the leaf. It is altogether a plant worth watching.

Cardinal Flower

Often you can locate a river by looking for a row of cottonwoods or sycamores. When we arrived in Saratoga County some thirty-seven years ago, I could trace the course of streams through pastures and meadows by the brilliant red of cardinal flowers (*Lobelia cardinalis*). This plant was sent home from America by French missionaries during Cardinal Richelieu's reign; the name is an allusion to the red robes of the office. The water still exists in some of these spots but hardly any cardinal flowers. Cardinal flower does not have rare plant status in New York, but was on the 1974 protected native plant list because such a conspicuous plant was likely to become rare from overpicking.

Cardinal flower has a wide distribution, from New Brunswick to Florida, and west to Texas and Colorado. Last summer I was hunting for a plant to observe frequently. On a backpack trip in August we met a single flower spike on the shore of the Oswegatchie just below thundering High Falls. Unfortunately, I could not visit there often. So the one plant in our yard, which prospered due to excessive rain, served as my model. My home specimen continued blooming through three frosts and showed no response to rain. (Many flowers close or change their orientation, apparently to keep their pollen dry.) Bloom progressed upwards on the spike. Eventually there were thirty-five flowers.

cardinal flower

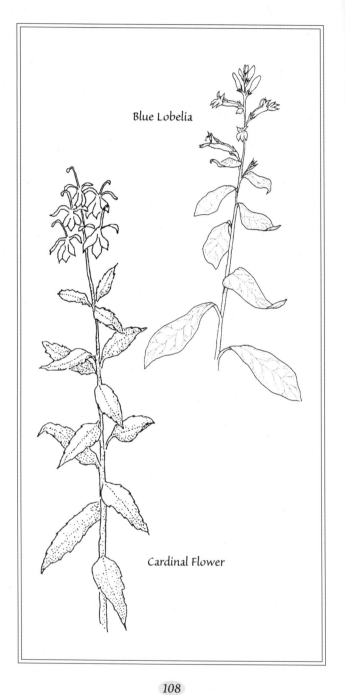

Blue Lobelia

Cardinal Flower

I labeled several flowers with colored twist-ties so I could follow their individual histories. Some twenty-six flowers began development to fruits, but only three matured.

Each flower has a calyx with five very thin sepals, a corolla with two lips—two lobes on the upper lip reaching out like arms raised at shoulder level and three spreading lobes on the lower lip. If you pull on this lip you'll notice a split down the back of the corolla tube. It opens like a hospital gown. The five stamens are united and wrapped around the pistil. All flower parts are attached to the top of the ovary. This type of anatomy, with bilateral symmetry and an inferior ovary (an ovary located below other flower parts), marks the plant as evolutionarily advanced.

There are thirty species of lobelias in the United States. Lobelias are placed in a family of their own or in the bluebell family. The name comes from Mallias de l'Obel, a Flemish physician of Elizabethan times.

blue lobelia

Next to my cardinal flower grows a sturdy cluster of great blue lobelia (*Lobelia siphilitica*). I watched many bumblebees fly in and out of these flowers. The space between stamens and pistil and the lower lip landing platform is much narrower in great blue lobelias than in the cardinal flower. None of these bumblebees visited the red flowers, but remained true to blue. Because the stamens of a lobelia flower mature before the pistil, self-pollination is not possible; so some agent must have accomplished the transfer of pollen on my successful cardinal flowers, those that became fruits. The literature suggests bumblebees, hummingbirds and day-flying moths. Pink and white cardinal flowers have been noted in the literature. It is easy to hybridize the cardinal flower and the great blue lobelia artificially. The hybrid has a pink corolla. Why do they not hybridize naturally since they share habitat and blooming time? One researcher tied two spikes together—red at bottom, blue at top. Visiting bumblebees began at the bottom, flying in and out of red flowers. They are said to have "buzzed angrily" when they reached the blue flowers and departed. It seems we have separate species thanks to the behavior of pollinators.

In late October, mature cardinal flower fruits still held on to the wilted flower remains at their tips but finally shed them and opened from the top of the capsules. Inside were many tiny dustlike seeds. It is likely that, being so small, the seeds lack stored food and need to rely on certain fungi to help them get established. This situation is well documented in orchids and other plants.

The cardinal flower is a perennial. Rhizomes spread out from the parent plant and produce rosettes of leaves that may send up flowering stalks the following year.

Fall

People from elsewhere envy us our foliage feast at this time of year. This is a good time to begin your acquaintance with individual woody plants. While admiring leaf colors you may want to learn the characteristic shapes of leaves of different species. Also check for fruit.

In addition to the white snakeroot mentioned in the following pages, there are many other members of the Composite family that bloom into fall. Sunflowers, asters and goldenrods are the best known.

Keying species is a challenge. It might be equally satisfying to watch the insects that visit these late bloomers for their last suppers and the spiders that prey on these insects. Along with gentians, described herein, we often find gerardia and the orchid ladies' tresses. Where mowing machines have done their deed we often find a second bloom of summer flowers. Enjoy them along bike trails and roads, and while you're at it don't forget to look for mushrooms. If there is enough rain, mushrooms can brighten the fall forest floor just as flowers did in the spring.

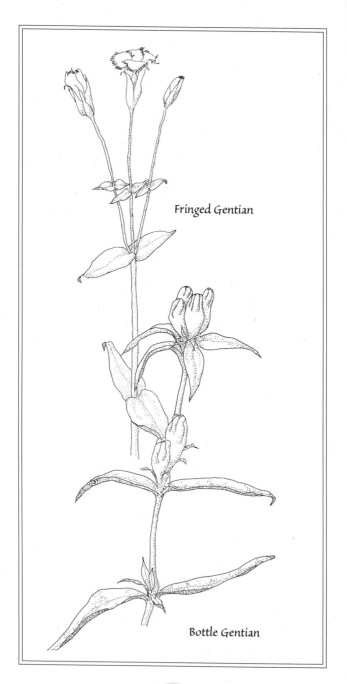

Fringed Gentian

Bottle Gentian

Fringed Gentian

A field of fringed gentian (*Gentiana* or *Gentianopsis crinita*) in bloom is a sight that inspires pilgrimages and challenges poets and photographers. Specialized habitat requirement is undoubtedly one reason for its rare status.

I am acquainted with four areas where fringed gentian grows, all of which have experienced human disturbance or fire: the soggy edges of the trail on Rocky Peak Ridge, in the High Peaks, where it grows with the more common bottle gentian; a field with calcareous seepage along the Vermont border in Washington County; Letchworth State Park; and an area in Wilton, Saratoga County. At the latter location, Orra Phelps, ADK's first resident naturalist, had sold sand from her property and created an ideal habitat, a combination of wetness held by clay and topped with sand. Orra's niece, Mary Arakelian, has documented this landscape engineering. The seeds are very small and light and could have arrived by wind. Fringed gentian has thrived in this field all these years in the company of ladies' tresses, grass of Parnassus, blue curls and bog clubmoss.

Seeds from these plants were scattered elsewhere by Orra and given to us pilgrims who came every September 10 to celebrate her birthday and the gentians. (None of our seeds succeeded.) Orra knew that the plants are biennials. She could identify the little seedlings that grow the first season and predict the flower crop for the following year.

It seems to me that over thirty years there must have been other introductions of seeds, or that some exceptional plants took on an annual habit or waited three years to bloom, or that some seed was longer-lived, surviving more than a year before germinating. By some means or other it has happened that seedlings and mature plants co-exist and every year is a blooming year. Blooming occurs from August to October and varies greatly in quantity.

A Cornell doctoral candidate discovered that while gentians produce plenty of seedlings, the tiny first-year plants are shaded out. Orra must have known that too. She pruned some of the invading woody plants and the good stewards of her land are continuing to do so.

In a 1991 publication of the Rowe Historical Society, *Wildflowers of Rowe, Massachusetts*, Susan Alix Williams remarks that fringed gentians are somewhat specific in their requirements, needing high levels of magnesium in the soil and a nearly neutral pH. She also warns that if a field is mowed before the seeds are set, the plants will disappear in that location. That is the unfortunate fate of plants that rely entirely on sexual reproduction with seeds that don't usually remain dormant past one winter.

Here are some of my fringed gentian experiences at Wilton:

9-13-90:

The sessile, opposite leaves are smooth, waxy and shiny, as is the stalk; two opposite sepals are wider than the other pair, so the cross section is really a rectangle, not a square. In each sepal the center vein forms a keel. The seams that join sepals are flanged. Only a few flower buds are open. They are two inches long.

8-28-91:

Prime time; noon on a sunny day (gentians are open only in sun). I can watch progression from

tight bud to senile flower.
A rectangular box of green sepals
pointed at the lip begins to split,
revealing a whirl of four fringed
blue petals.

The box opens more. The
petals unfurl. The bluest blue is
at the top of the fringed petals
and runs down the veins. The
flower is just short of wide
open and the two lobes of the
stigma are closed like lips
pressed together.

Four anthers are shedding yellow pollen. They
are attached at a jaunty angle to four sturdy
filaments that spring from the petals. Interesting
that the anthers are level with the ovary and are
below the stigma. Self-pollination seems to be
discouraged both by the timing of pollen release
and by stigma receptivity and placement of
organs. Open flowers have lush yellow-green
stigmas and empty anthers. A matching yellow-
green spider sits in one flower. A small bee
emerges from another—a pollinator, I hope.

There are hundreds of blooms around me.
Plants vary greatly in size, from one to three feet
tall, and in the number of branches. The cham-
pion in the little population surrounding me
bears twenty-two buds and flowers. At Letchworth State Park, forty to fifty blooms were
counted on plants. I stick my nose in various
flowers—no smell, at least none at high noon.

Because fringed gentians have four sepals and four
petals, the flowers are called four-merous. The bottle
gentian, and most other gentians, are five-merous.
Gentian flowers come in various colors—white, green,
yellow—in addition to the heavenly blue.

The rhizome of the yellow gentian of European mountains is the source of a medicine. A bitter spring tonic was made from native gentians by our early settlers. It was often enriched with gin or brandy. It is listed among medicinal Shaker herbs and came well recommended from Roman times. The name *gentian* came from Centius, an Illyrian king who cured a mysterious fever that had stricken his army. I find this flower to be a good visual cure for whatever ails me.

September Fungi

Late September found three people and a dog back packing north from Piseco along the Northville–Placid Trail. With flowers gone, except for an occasional aster or dalibarda, color was supplied by maple leaves and fungi.*

Almost every cut log edging the trail sprouted a bear's head toothed fungus, *Hericium*. Maybe the early bear season and the hunter's question, "Have you seen a wounded bear," influenced me in my identification. It could have been a different species of *Hericium*. Some species look like coral, some are covered with "ice-like spines" others with "ice-like white teeth." Unwilling to carry the *Hericium* in hand all day or mash it in the pack, we left it to spread its spores and decay.

By afternoon the dog cast off his pack, his mistress added five pounds of dog food to her pack, and I acquired a mushroom collecting bag.

We passed honey agarics clustered around tree trunks, tons of honey agarics, many too old, but some still in their prime. This mushroom grows in clusters on wood of all kinds. It has hairs on the center of a yellow-brown cap and a persistent ring around the central stem; it makes a white spore print. It also produces runners that look like black shoelaces between the bark and wood of affected trees. These can

* Fungi have been ousted from the kingdom of plants and now occupy a kingdom of their own, but I trust the reader will not be perturbed by the inclusion of this essay.

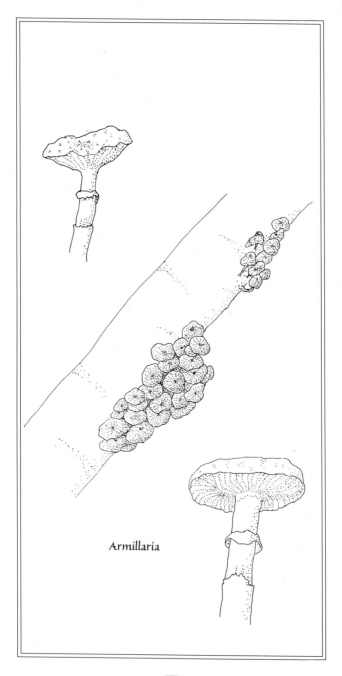

Armillaria

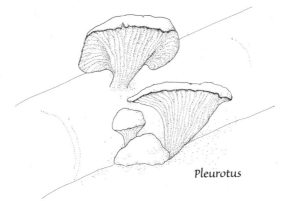

Pleurotus

travel long distances, infecting other trees.

You will find this honey agaric, honey mushroom or shoestring fungus in any book on fungi under *Armillariella mellea* or *Armillaria mellea*, a demonstration that fungi's scientific names are often quite changeable. In Italian neighborhoods you can buy the honey agaric canned as Chiodini al Naturale. Miller (see Selected References) considers all members of this genus—*Armillaria* or *Armillariella*—excellent eating. Other writers purposely omit the honey agaric from foraging lists because beginners might confuse it with deadly *Galerinas*. Today's mushroom field guides all contain warnings; consult several sources of information, including an experienced mycologist, before eating wild mushrooms. Moral: learn to observe carefully and enjoy safely.

I cut off choice specimens for my bag. At Spruce Lake we sliced them, cooked them well in a little margarine squeezed out of a Gerry tube and feasted. My companions ate rather sparsely since experts warn us to eat only small amounts of new mushrooms. On a backpack trip especially one ought to heed warnings.

Next day and thereafter rain was our companion. Eyes downcast I reveled in the beauty of oyster mushrooms of various kinds. Angel wing (*Pleurotus porrigens*) shone whitely on mossy logs. It is a small (one to three inches across), fragile oyster mushroom with white spores. I cut one batch of *Pleurotus sapidus* or *Pleurotus ostreatus* from a dead tree. That night at Cedar Lake we cooked the water-soaked fungus in a little margarine. This time we got a dish resembling oyster soup. I liked it, but my friends only tasted it.

The mushroom's common name could be derived from its flavor and texture as well as its shape. Some books suggest cooking it in butter with onion and garlic, adding evaporated milk, seasonings, and even real oysters. Our oyster mushroom consisted of clustered, soft, pliant, whitish, fan-shaped caps, about two to six inches across, attached to a dead tree. It also exists on living trees.

The two species of *Pleurotus* are lumped together in some books, separated in others by their respectively white (*ostreatus*) and lilac (*sapidus*) spores. The mushroom fruits year after year in the same place. In rainy years, it may fruit repeatedly. For those who enjoy it, a location map for *Pleurotus* comes in handy.

Next day I cast my eyes on *Pholiotas* with shiny, viscid, spotted caps. Among those that had lost their spots or collapsed from too much water there were some scrumptious looking ones. During the summer I had eaten one *Pholiota* (of doubtful species) on another Adirondack trip and thoroughly enjoyed it. But I had no book to consult on this trip and therefore desisted. Fall is a particularly good time to enjoy mushrooms visually and tactily. If you want to eat them too, learn thoroughly and experiment carefully one kind at a time.

White Snakeroot

During mushrooming season, September and October, white snakeroot (*Eupatorium rugosum*) provides the flower accent. A fellow mushroomer thought the flower clusters resembled cauliflower. Close up you will see pistils and stamens protruding from the clusters.

To help you understand the flower structure let's look at the Composite family, to which white snakeroot belongs. This enormous family is known for flowers grouped into tight clusters, or heads. We can divide it into categories according to the types of flowers found in the heads: all strap-shaped flowers, like dandelion; all tube-shaped flowers, like white snakeroot; both kinds of flowers, the strap-shaped ones forming an outer circle, the rays, and the tube-shaped ones the center, like sunflower.

Each head of white snakeroot has eight to thirty perfect tube flowers. What makes a flower perfect? Having both male and female functioning sex organs. Each flower has five bright white fused petals, one pistil and five stamens. Use a hand lens to appreciate fully the beauty of the individual flower. Then look at the green bracts surrounding each bouquet of flowers. You'll find a row of them, linear, slightly unequal, scarcely overlapping.

On October 3, I labeled a flower head in full bloom (with a colorful wire twist-tie). While I was thus engaged, a bumblebee paid a visit to the flowers. I watched this cluster as the bloom closed, faded and shriveled.

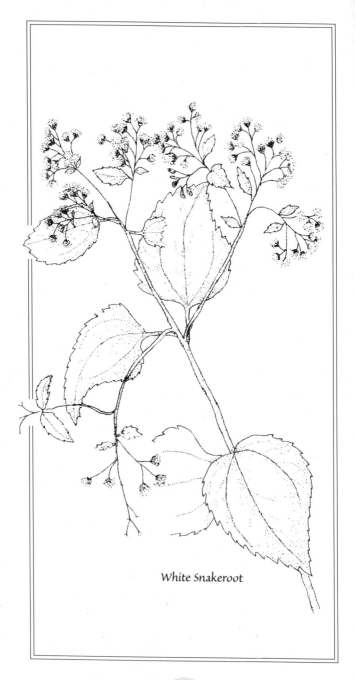

White Snakeroot

A month later, on Election Day, came the reopening into full fuzzy fruit heads. Each fruit is just over one-sixteenth inch long, dark and ribbed. Attached to its top is a parachute of about fifty tawny hairs. A few days after the reopening, fruits began to depart with the breeze.

White snakeroot is a perennial, growing in woods, especially edges of woods, in most of the eastern half of the United States. New shoots grow from the underground overwintering stem. As I write this, in mid-May, the shoots have just appeared above ground. In favorable habitats, the plants will be five feet high by September. The stems are smooth, often purple in the older parts, and tend to have a rectangular cross section. The leaves are opposite, three-veined and coarsely toothed.

If you look for snakeroot in wildflower books, you will find many unrelated ones. In every locality some plant had a medicinal reputation as an antidote to snake bite and was give the name "snakeroot." Our snakeroot, *Eupatorium rugosum*, was named for a King of Pontus, Mithridates Eupator. The species name "rugosum" means wrinkled. I presume this refers to the protruding, "varicose" veins on the underside of the leaf.

During the early years of the United States, a dreaded disease called milk sickness often reached epidemic proportions. Weakness, nausea, vomiting, constipation, the odor of acetone on the breath and muscular tremors were among the symptoms. Many, including Abraham Lincoln's mother, died; others remained incapacitated for life. Dr. John Kingsbury, in *Deadly Harvest*, tells us that during the first half of the nineteenth century, milk sickness was so devastating that in some areas it reduced the population to less than half in a year or two. The disease was associated with certain locations and with cows that suffered from "trembles," an appropriately named condition. Poison ivy was blamed, as were miasmas from certain soils, even spider webs, and after Pasteur, bacteria.

It took a long time to establish the link between white snakeroot, grazing cows, and humans ill with milk sickness. Trembles develop very slowly. The toxic factor is fat-soluble and is concentrated in the milk of the lactating cow. Humans or calves drinking milk from the family cow that has been turned out in the woodlot could thus develop symptoms long before the cow itself. Because tremetol, the toxin, is excreted in milk, milking animals themselves are afforded some protection. This knowledge has inspired a hare-brained scheme of ridding an individual of fat-soluble toxins by hormonally inducing lactation. The general public no longer needs to worry about white snake-root and the safety of milk, but the self-sufficient homesteader needs to be aware of the danger.

The genus *Eupatorium* contains many species. Locally, in wetlands, we have several Joe Pye weeds and bonesets, all used medicinally by Native Americans. But there is only one seriously toxic member—white snakeroot.

Winter

If you became acquainted with trees and shrubs while they were in leaf, now is your chance to get to know them better by examining buds, leaf scars and bark. I like to guess the identity of a tree from a distance by its shape, branching pattern and location. Then I confirm or correct with a closer look.

Now the details of woody plants stand out. At the same time we can see farther and absorb a larger picture. If it is too cold to use a key or even to stop, you can enjoy reflecting on the composition of the forest: how it changes with latitude, altitude, the aspect of its slope, and with natural and human disturbances.

This is a good time to use a hand lens to look at mosses, liverworts, and lichens growing on tree trunks. When the snow melts, other mosses show up as intense green patches on the ground. Evergreen ferns and club mosses stand out, as do flowering evergreen plants such as pipsissewa, pyrola and those mentioned in the essay on wintergreens.

You may enjoy pruning woody plants in late winter and bringing twigs indoors to watch bud opening. (Keep the twigs in water.) Before the vernal equinox, elms, silver maples, alders, coltsfoot and skunk cabbage will open their flower buds and announce spring.

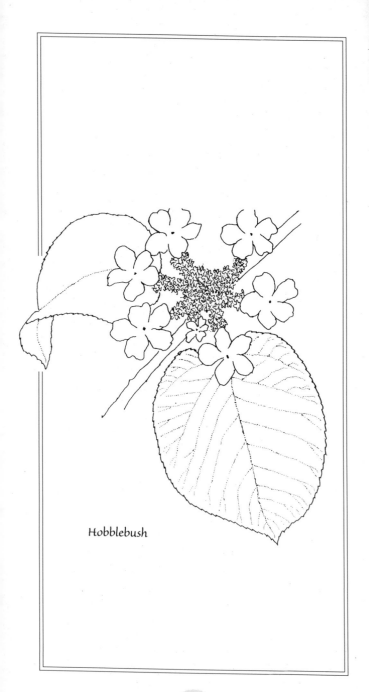

Hobblebush

A Bush for
All Seasons:
Hobblebush

I n the midst of winter there is one woody plant,
hobblebush (*Viburnum alnifolium* or *V. lantanoides*),
that more than any other holds the promise of
spring. "We'll open tomorrow" the buds seem to
shout. Two thick, fuzz-covered, tightly folded leaves,
each side of the leaf curled into the center—this is the
one-half- to three-quarter-inch bud from which
growth will unfold in May.

Because there are no bud scales, it is called a naked
bud. I can only think of two other native shrubs with
naked buds: witch hazel and poison ivy. The buds are
paired at the tips of branches. Between the pair may be
another growth bud promising a huge flower cluster.
This flat cluster is five to six inches across. It consists of
several branches, each of which subdivides. Each stem,
except the central one, presents one or two showy
flowers on a lengthened stalk. These are impressive,
but sterile, flowers—one inch wide, tubular with five
unequal, spreading lobes—the larger ones directed
towards the outside.

There are no sex organs. The white or pinkish
flowers are merely advertisement. (Another wild
viburnum, highbush cranberry, has similar sterile
outer flowers, and a cultivated viburnum, the snowball
bush, has nothing but sterile flowers.) On Camel Mt.

in the Dix Wilderness area, at about 2500 feet, the flags were inviting guests on May 14, prematurely.

When will the feast follow the invitation? Not until the clusters open into the real flowers. Like all viburnums, these flowers have five joined sepals, five joined white petals, five stamens and three tiny stigmas at the tip of the ovary.

As I watched my flowers open, the stamens became visible and a few days later I could finally see a three-pronged stigma in the center. As in many plants, male and female organs mature at different times, encouraging cross-pollination.

One gets so involved with the handsome flowers, one hardly pays attention to the leaves as they unroll and expand. They become four to eight inches long, heart-shaped at base, veiny, finely toothed at margin. The leaf stems, lower leaf surface and young branchlets are still coated with cinnamon brown, stellate hairs. (Use a lens to see the branched hairs.) We presume hairiness protects from dessication. It may also repel some ticklish caterpillars.

The twig bark is shiny olive brown to purple brown; sometimes there is a grayish film. Lenticels—the irregularly spaced, raised spots on the bark through which gas exchange takes place—are light brown.

In late summer hobblebush leaf color announces fall—orange to deep maroon, often in two different patterns on each side of the center vein. It would be fun to give students in a painting class such a leaf to see if they could match the hues. When the leaves drop they leave behind three-lobed leaf scars with three bundle scars. The latter scars mark the spot where vascular bundles entered the leaf as they transported food and water between branch and leaf.

Meanwhile, the fruit is ripening. At one time you will see green, orange, red, purple and black fleshy fruits, about one-third of an inch long. Inside each is one flat seed. I tasted the nontoxic fruits. Red fruit is astringent; purple-black is bland and pleasant. Ruffed grouse like it, as do other animals. Do they spit out the seed, or "plant" it after it goes through their gut?

Hobblebush is also known as moosewood, a name it shares with striped maple, its frequent companion in our woods. When I did library research on deer browse, I found that hobblebush is listed by several authors as among the most preferred, and probably most nutritious, browse, especially twigs no thicker than wooden matchsticks. When deer are very hungry they will eat browse the size of wood pencils. These have a smaller proportion of protein-rich young bark. If deer are forced to eat frozen twigs, they use much energy to thaw them out. I learned of research in a Pennsylvania forest that determined deer browsing on hobblebush was so extensive that the plant was virtually exterminated. I also learned that deer defecate on the average thirteen times every twenty-four hours. I can now correct my overestimation of deer population as judged by manure piles on ski trails.

It is my impression that hobblebush is the most common Adirondack shrub. My next impression is that deer shape the shrub. The ten foot height is rarely reached where there are deer. The branching pattern, the richness of bloom and subsequent fruit depend much on browsing.

A third common name for our shrub is witch hobble. Why the word *hobble*? In a good stand of hobblebush, follow low, horizontal or drooping branches. With luck you will find some that have rooted at the tip. This is a good way to spread over rocky terrain to the next bit of hospitable soil in a crack. Is the shrub hobbling or do we hobble after being tripped by these branches? The illustrator thinks the bush acts as a hobble for us, slowing our progress, just like the hobbles that are put on horses to keep them from straying.

Fir

Scotch Pine

Spruce

Christmas Trees

At holiday season we leave the trail for a while to check on trees available at Christmas tree plantations, then return to our own forests to look at potential Christmas trees.

The Saratoga Tree Nursery[1] supplies seedlings of conifers, as well as some hardwoods, shrubs and vines. Any New York State landowner may purchase these seedlings as long as they will not be used for ornamental or landscape purposes.

If you qualify, you can order seedlings in units of 100, which will cost you $25.00. What you will receive in early spring is a bag of two- or three-year-old seedlings, about six to twelve inches tall, with recommendations for spacing, "enough distance between rows to accommodate two-foot to three-foot growth plus the width of mowing equipment."

As I read the information leaflet I had to laugh when I came to the statement, "The most common cause of seedling mortality is mowing." I once participated in a big wildlife planting at a school. Trees and shrubs were planted with care. Bucket brigades supplied water. A couple of weeks later the groundskeeper did a neat mowing job, decapitating all. If you guard your seedlings carefully, they will reach Christmas tree size in six to twelve years.

Of the available conifers, eight are listed as Christmas trees:

1. Saratoga Tree Nursery, New York State Department of Environmental Conservation, 431 Route 50S, Saratoga Springs, NY 12866

- Balsam fir, red pine, white pine, white spruce—native to New York
- Douglas fir—from western United States
- Austrian pine, Norway spruce, Scotch pine—from Europe

Now a look at conifers in our woods. We have a woodsy acre-and-a-quarter yard with volunteer conifer seedlings. Maybe you too have access to a potential Christmas tree. I've just browsed through E. H. Ketchledge's *Forests & Trees of the Adirondack High Peaks Region* (see Selected References) with my Christmas tree search in mind. Northern white cedar appeals to me, especially its smell, but no one considers it a candidate. Tamarack, or larch, sheds its needles long before December, and hemlock drops needles even before you get it set up inside.

That leaves, as potential candidates, spruces, firs and pines. All three belong to the Pine family. Once I was engaged by a garden club to give a slide show at a member's house—"the house with the big pine tree in front." I cruised up and down the block; not a single pine there. I finally knocked on a door behind a spruce and learned to interpret "pine" more liberally.

How do you differentiate among members of the Pine family? *P* is for pine and package—our Eastern pines have needles that come in bundles of two, three or five wrapped together by a sheath. *S* is for spruce and square—the needles are four-angled, making a square- or diamond-shaped cross section. *F* is for fir and flat—the needles are so flat they cannot be rolled between the fingers; they are attached directly to the twig, without a stem; and when they drop off they leave a smooth, circular scar.

Spruces have a very appealing shape. White spruce bears many attractive cones at a young age. Our children always picked that naturally decorated tree on a tree farm, even though it lost many cones as it was dragged to the car. The fact that it drops needles faster than firs or pines did not matter as it was cut a day or two before Christmas. The fact that it often had a skunky odor bothered only my nose.

I wondered why our common red spruce, although "frequently used as a Christmas tree" according to William Chapman and Alan Bessette (*Trees and Shrubs of the Adirondacks: A Field Guide*), was not in the recommended list or available as seedling. The answers came from Ed Ketchledge. I'll summarize: it is slow growing; the short needles are dull green and cluster around the twigs, giving the tree a bare appearance; and the needles drop even faster than those of other spruces.

Balsam firs tend to be symmetrical with attractive green, soft and fragrant needles. They should be the favorite species, but the highest marks go to Austrian and Scotch pines.[2] Indeed, Scotch pine, grown under controlled conditions, is the leading Christmas tree in both New York and the whole country.

2. "Selection and Care of Your Christmas Tree," Information Bulletin 48, includes a chart describing characteristics of common Christmas trees under room conditions. The bulletin is available through Cooperative Extension.

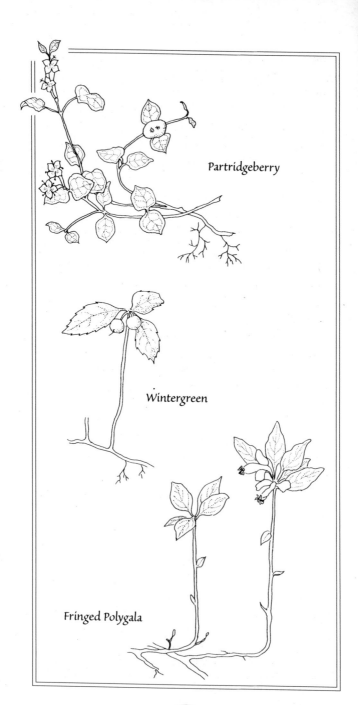

Partridgeberry

Wintergreen

Fringed Polygala

Wintergreens

As the snow melts, our eyes rejoice in patches of green ground cover. There are mosses and liverworts, club mosses and spikemosses, evergreen ferns flattened by winter's burden of snow, ground yew and some evergreen broad-leaved flowering plants. Here we shall get better acquainted with three native members of that group: partridgeberry, fringed polygala (also known as gaywings, flowering wintergreen) and wintergreen (or teaberry, checkerberry).

Partridgeberry (*Mitchella repens*, in the Bedstraw family) is a trailing vine, hugging the ground, often so densely leafed that the stem is not visible. The leaves are opposite, round, about one-half inch long, with a conspicuous light center line. Many people confuse wintergreen (*Gaultheria procumbens*, in the Heath family) and fringed polygala (*Polygala paucifolia*, in the Milkwort family) when they are not blooming. Both are upright plants rising from primary stems that are hidden in the soil. The general shape may be similar, but fragrance, flavor and stiff, shiny, waxed texture belong only to wintergreen. Fringed polygala leaves are dull green, often with a tint of purple and they appear more crowded. To honor the similarity and confuse us, the name "flowering wintergreen" has been bestowed on fringed polygala. If you look carefully, you will notice small scale-like leaves on the lower stem of fringed polygala, but not on wintergreen.

The flowers of fringed polygala, superficially orchid-like or airplane-like, are indeed showy. One to four rose-purple to white flowers are born near the top of the plant. (White is rare. I have seen only one white-flowering plant, at Barberville Falls, a Nature Conservancy preserve in Rensselaer County.) Two lateral sepals make the wings. The three petals form a tube with a fringed crest. My botany book tells me that the flower ripens into a fruit which is a "two-seeded capsule that dehisces along the margin." I have never seen one. It will be a quest this summer. There are also insignificant-looking small, never-opening, self-fertilizing flowers below the ground.

While fringed polygala blooms in spring like most of our woodland plants, wintergreen and partridge-berry are usually summer bloomers. Partridgeberry has white-tubed twin flowers. The four lobes are fringed on the inside. The flowers are either male with abortive pistils or female with abortive stamens. The ovaries of paired flowers fuse to form a red, berry-like, edible fruit. Fruits persist into the following summer, giving us many chances to examine the eyes, the remains of two flowers on each fruit.

Partridgeberry is an excellent plant for terrariums. You need only take a cutting from a wild plant, leaving roots intact. Your cutting will root at the joints, grow in the dim light of an indoor terrarium, and bloom in early spring.

Wintergreen bears one to three nodding, bell-shaped flowers with five lobes. The calyx, which surrounds the tubular corolla, becomes the capsule, forming a berry-like fruit. This edible fruit, like the leaves, smells and tastes like tooth-paste. The same refreshing substance occurs in twigs of yellow and black birches. Orra Phelps pointed out to me that the fruit increases in size during the winter.

Besides the evergreen habit, shared home and edible fruits, wintergreen and partridgeberry have another common denominator. Both have generic names derived from men who thus gained immortality. Gaulthier was a Canadian physician, host, and guide to Peter Kalm, a young Swedish botanist who collected plants for the man who named and classified them—Linnaeus. Mitchell, a Virginia botanist and correspondent of Linnaeus, brought American plants to Europe. The ship was looted. He arrived empty-handed but was able to draw a map of America that was used by King George III when he agreed to American peace terms in 1783.

Selected References

Throughout this book you will find references to books and journals. I have listed a few especially useful ones here.

NATURE STUDY HANDBOOKS

Comstock, Anna Botsford. *Handbook of Nature Study*. Ithaca, N.Y.: Cornell University Press, 1986. Contains useful information about all phases of nature and suggested observations and activities. Written for teachers.

Gibbons, Euell. *Stalking the Healthful Herbs*. New York: David McKay Co., 1966. This author may tempt you to eat the weeds.

Kudish, Michael. *Adirondack Upland Flora*. Fleischmanns, N.Y.: Purple Mountain Press, 1992. A wonderful book full of information about plants and environmental factors that influence plants.

Palmer, E. Laurence and H. Seymour Fowler. *Fieldbook of Natural History*. 2d ed. New York: McGraw-Hill Co., 1975. Broad coverage of topics from rocks to fungi, plants, and animals. Well illustrated. This book is now out of print and outdated, but it is an excellent guide that may be found in a library or used bookstore.

Pepi, David. *Thoreau's Method*. Englewood Cliffs, N.J.: Prentice-Hall, Inc., 1985. Hints on appreciating nature, go-light equipment, excursions, activities, thought-provoking questions, and recommended reading. This book is now out of print, but worth locating in a library or used bookstore.

Serrao, John. *Nature's Events: A Notebook of the Unfolding Seasons*. Mechanicsburg, Pa.: Stackpole Books, 1992. Beautifully written personal observations of plant and animal life and accurate reports from the scientific literature.

Stokes, Donald W. *Nature Guides*. Boston: Little, Brown. A series of books on wild shrubs and vines, wildflowers, birds, insects, etc., emphasizing observation and reporting on research.

FIELD GUIDES

If you use a guide that is geographically restricted, you will find more local species in its pages and be frustrated less frequently.

Bessette, Alan. *Guide to Some Edible and Poisonous Mushrooms of New York*. Utica, N.Y.: North Country Books, 1988.

Bessette, Alan. *Mushrooms of Northeastern North America*. Syracuse, N.Y.: Syracuse University Press, 1996.

Graves, Arthur H. *Illustrated Guide to Trees and Shrubs: A Handbook of the Woody Plants of the Northeastern United States and Adjacent Canada*. Mineola, N.Y.: Dover Publications Inc., 1992.

Hall, Alan. *The Wild Food Trailguide*. Orlando, Fla.: Holt, Rinehart and Winston, Inc., 1973. Author is an upstate New Yorker and a Cornell graduate. This work is more concise and trustworthy than most of the dozens of books published more recently on wild foods.

Ketchledge, E. H. *Forests & Trees of the Adirondack High Peaks Region: A Hiker's Guide*. Lake George, N.Y.: Adirondack Mountain Club, 1996.

Lincoff, Gary A. *The Audubon Society Field Guide to North American Mushrooms*. New York: Alfred A. Knopf, 1981.

McGrath, Anne. *Wildflowers of the Adirondacks*. Utica, N.Y.: North Country Books, 1981.

Miller, Orson K. *Mushrooms of North America*, New York: Dutton, 1978.

Newcomb, Lawrence. *Newcomb's Wildflower Guide*. New York: Little, Brown, 1977. An ingenious new key system for quick, positive field identification of wildflowers, flowering shrubs, and vines.

Peterson, Roger Tory. *A Field Guide to Wildflowers of Northeastern and Northcentral North America*. Boston: Houghton Mifflin Co., 1975.

Phillips, Roger. *Mushrooms of North America: The Most Comprehensive Mushroom Guide Ever*. Boston: Little, Brown, & Co., 1991.

Slack, Nancy and Allison Bell. *85 Acres: A Field Guide to the Adirondack Alpine Summits*. Lake George, N.Y.: Adirondack Mountain Club, 1993.

SCIENTIFIC REFERENCES

Gleason, Henry and Arthur Cronquist. *Manual of Vascular Plants of Northeastern United States and Adjacent Canada*, 2nd ed. New York: New York Botanical Garden, 1991. The scientific names used in *Trailside Notes* agree with those used by Gleason and Cronquist.

OF GENERAL INTEREST

Busch, Phyllis. *Wildflowers and the Stories behind Their Names*. New York: Charles Scribners & Sons, 1977. A well-written book, beautifully illustrated by Anne Ophelia Dowden. Look at any and all books by this wonderful illustrator. This book is now out of print, but worth locating in a library or used bookstore.

Carson, Rachel. *The Sense of Wonder*. New York: Harper and Row, 1956. This will lead you to read other books by the author.

Hubbel, Sue. *A Country Year: Living the Questions*. New York: Random House, 1986. See also her work, *A Book of Bees*.

Johnson, Charles W. *Bogs of the Northeast*. Hanover and London: University Press of New England, 1985. A readable discussion on bogs and fens, including descriptions of carniverous and other plants that grow there.

Russell, Helen Ross. *Foraging for Dinner*. Nashville, Tenn.: Thomas Nelson, 1975. This is a great introduction to botany through eating. The author is an excellent teacher whose workshops and other books I have enjoyed thoroughly. This book is now out of print, but worth locating in a library or used bookstore.

Index

About the Author

Ruth Schottman is an environmental educator well known and respected for her field and book knowledge and the many years of experience she brings to both. She teaches courses and workshops for the Adirondack Mountain Club (ADK), and the Environmental Clearinghouse of Schenectady (ECOS). In addition, she leads wildflower walks for the Saratoga Land Conservancy, the Saratoga Springs Open Space Project, and ECOS. Schottman holds a bachelor of science in genetics from Cornell University. She lives in Burnt Hills, New York.

About the Illustrator

Clem Habetler's Trailside Notes illustrations have graced the pages of *Adirondac* for sixteen years. Her botanical illustrations have also appeared in *Wildflowers Along the Way*, published by ECOS in 1989. An avid reader of environmental literature and former newsletter editor for ECOS, Habetler holds a bachelor of science in chemistry from Carnegie-Mellon. She is a resident of Rexford, New York.

Your Trailside Notes

TRAILSIDE NOTES

Join Us!

We are a nonprofit membership organization that brings together people with interests in recreation, conservation, and environmental education in the New York State Forest Preserve.

MEMBERSHIP REWARDS

Discovery
ADK can broaden your horizons by introducing you to new people, new places, recreational activities and interests.

Member Benefits
- 20% discount on all ADK publications, including guidebooks and maps
- 10% discount on ADK lodging facilities in the Adirondacks
- 10% discount on ADK logo merchandise
- Reduced rates for educational programs
- One-year subscription to *Adirondac* magazine
- Membership in one of ADK's 26 chapters
- Member-only outings to exciting destinations around the world

Satisfaction
Knowing you're doing your part to protect and pre-serve our mountains, rivers, forests and lakes to ensure that future generations will be able to enjoy the wilderness as we have.

Join a Chapter

Three-quarters of ADK members belong to a local chapter. Those not wishing to join a particular chapter may join ADK as members-at-large.

Chapter membership brings you the fun of outings and social activities and the reward of working on trails, conservation, and education projects at the local level. Chapter membership is included in your benefits.

Adirondak Loj .... North Elba
Albany .... Albany
Algonquin .... Plattsburgh
Black River .... Watertown
Cold River .... Long Lake
Finger Lakes .... Ithaca–Elmira
Genesee Valley .... Rochester
Glens Falls .... Glens Falls
Hurricane Mountain .... Keene
Iroquois .... Utica
Keene Valley .... Keene Valley
Knickerbocker .... New York City and vicinity
Lake Placid .... Lake Placid
Laurentian .... Canton-Potsdam
Long Island .... Long Island
Mid-Hudson .... Dutchess Co.
Mohican .... Westchester and Putnam counties, NY/
                                                                Fairfield Co., CT
New York .... Metropolitan Area*
Niagara Frontier .... Buffalo
North Jersey .... Bergen County
North Woods .... Saranac Lake–Tupper Lake
Onondaga .... Syracuse
Ramapo .... Rockland & Orange counties
Schenectady .... Schenectady
Shatagee Woods .... Malone
Susquehanna .... Oneonta

Special requirements apply

Your membership is not limited to one chapter. You may affiliate with as many chapters as you like for a small fee (the cost of the newsletter).

Membership
To Join

Call **1-800-395-8080** or send this form with payment to

Adirondack Mountain Club
814 Goggins Rd., Lake George, NY 12845-4117.

Check Membership Level:

☐	Patron	$125*
☐	Supporting	$75*
☐	Family	$45*
☐	Adult	$40
☐	Senior Family	$35*
☐	Senior (65+)	$30
☐	Junior (under 18)	$25
☐	Student (18+, full time)	$25

School _____

*Includes associate/family members

Adirondack
ADK
Mountain Club

Name _____

Address _____

City _____ State _____ Zip _____

Home Telephone (_____) _____

☐ I want to join as a member-at-large.

☐ I want to join as a _____ Chapter member.

(For more information on Chapters, call 518-668-4447, ext. 30.)

List spouse & children under 18 with birthdates:

Spouse _____ Birthdate _____

Child _____ Birthdate _____

Child _____ Exp. date

Bill my: ☐ VISA ☐ MASTERCARD ☐ DISCOVER ☐ AM. EX

_____
Signature (required for charge)

ADK is a non-profit, tax-exempt organization. Membership fees are tax deductible, as allowed by law. Please allow 6–8 weeks for receipt of first issue of *Adirondac*.

All fees subject to change.

TSN

Backdoor to Backcountry

ADK outings and workshops— there's one just right for you!

ADK offers friendly outings for all skill levels—for those just getting started in local chapters, to Adirondack bushwhacks and international treks. Learn gradually through chapter outings or attend one of our schools, workshops, or other programs. A sampling includes:

- Alpine Flora
- Ice Climbing
- Rock Climbing
- Basic Canoeing
- Bicycle Touring
- Cross-country Skiing

- Mountain Photography
- Winter Mountaineering
- Birds of the Adirondacks
- Geology of the High Peaks
  ... and so much more!

For more about our workshops:

ADK Education Department
P.O. Box 867, Lake Placid, NY 12946
(518) 523-3441   9:00 a.m.–7:00 p.m.

For information about the Adirondacks or about ADK:

ADK's Information Center & Headquarters
814 Goggins Road, Lake George, NY 12845-4117
(518) 668-4447
Exit 21 off I-87 ("the Northway"), 9N South

May–Columbus Day: Mon.–Sat., 8:30 a.m.–5:00 p.m.
Tues. after Columbus Day–May: Mon.–Fri., 8:30 a.m.–4:30 p.m.

To join ADK by credit card, please call our toll-free number: 800-395-8080 (8:30 a.m.–5:00 p.m., M–F). Callers who join may take immediate discounts on ADK publications, workshops, ADK logo merchandise and lodge rates and may charge all to Visa, Mastercard, Discover or American Express.

For information about lodges, cabins, or campsites on ADK's Heart Lake property in the High Peaks region:

ADK Lodges
PO Box 867, Lake Placid, NY 12946
(518) 523-3441   9 a.m.–7:00 p.m.

Visit our web site: www.adk.org

The Adirondack Mountain Club, Inc.
814 Goggins Road
Lake George, NY 12845-4117
(518) 668-4447/Orders only: 800-395-8080 (M–F, 8:30–5:00)

BOOKS

85 Acres: A Field Guide to the Adirondack Alpine Summits

Adirondack Canoe Waters: North Flow
Adirondack Canoe Waters: South & West Flow

The Adirondack Mt. Club Canoe Guide
to Western & Central New York State

Adirondack Park Mountain Bike Preliminary Trail and Route Listing

The Adirondack Reader

Adirondack Wildguide (distributed by ADK)

An Adirondack Sampler I: Day Hikes for All Seasons

An Adirondack Sampler II: Backpacking Trips

Classic Adirondack Ski Tours

Climbing in the Adirondacks: A Guide to Rock & Ice Routes

Forests & Trees of the Adirondack High Peaks Region

Guide to Adirondack Trails: High Peaks Region

Guide to Adirondack Trails: Northern Region

Guide to Adirondack Trails: Central Region

Guide to Adirondack Trails: Northville–Placid Trail

Guide to Adirondack Trails: West-Central Region

Guide to Adirondack Trails: Eastern Region

Guide to Adirondack Trails: Southern Region

Guide to Catskill Trails

Kids on the Trail! Hiking with Children in the Adirondacks

Our Wilderness: How the People of New York Found,
Changed, and Preserved the Adirondacks

Trailside Notes: A Naturalist's Companion to Adirondack Plants

Winterwise: A Backpacker's Guide

With Wilderness at Heart:
A Short History of the Adirondack Mountain Club

MAPS

Trails of the Adirondack High Peaks Region
Trails of the Northern Region
Trails of the Central Region
Northville–Placid Trail
Trails of the West-Central Region
Trails of the Eastern Region
Trails of the Southern Region

Price list available on request.